The Runmakers

A NEW WAY TO RATE BASEBALL PLAYERS

The Runmakers

FREDERICK E. TAYLOR

The Johns Hopkins University Press BALTIMORE

The Johns Hopkins University Press
2715 North Charles Street
Baltimore, Maryland 21218-4363
www.press.jhu.edu

Library of Congress Cataloging-in-Publication Data

Taylor, Frederick E.
   The runmakers : a new way to rate baseball players / Frederick E. Taylor.
      p.   cm.
   Includes bibliographical references and index.
   ISBN-13: 978-1-4214-0010-5 (hardcover : alk. paper)
   ISBN-10: 1-4214-0010-3 (hardcover : alk. paper)
   1. Baseball players—Rating of. I. Title.
   GV865.A1T39 2011
   796.3570727—dc22

                              2010028479

A catalog record for this book is available from the British Library.

Special discounts are available for bulk purchases of this book. For
more information, please contact Special Sales at 410-516-6936 or
specialsales@press.jhu.edu.

The Johns Hopkins University Press uses environmentally friendly book
materials, including recycled text paper that is composed of at least 30
percent post-consumer waste, whenever possible. All of our book papers
are acid-free, and our jackets and covers are printed on paper with
recycled content.

To my wife
RUTH HAGER TAYLOR
Enabler, Supporter, and Partner

Contents

Preface ix
List of Abbreviations xi

PREGAME ANALYSIS 1

Part I. Every Era Has Its Greats

1. THE ERA OF CONSTANT CHANGE, 1876–1892
The Age of Dan Brouthers 23

2. THE LIVE BALL INTERVAL, 1893–1900
The Age of Ed Delahanty 38

3. THE DEAD BALL ERA, 1901–1920
The Age of Ty Cobb 49

4. THE LIVE BALL ERA, 1921–1941
The Age of Babe Ruth 61

5. THE LIVE BALL CONTINUED ERA, 1942–1962
The Age of Ted Williams 77

6. THE DEAD BALL INTERVAL, 1963–1976
The Age of Hank Aaron 93

7. THE LIVE BALL REVIVED ERA, 1977–1992
The Age of Mike Schmidt 106

8. THE LIVE BALL ENHANCED ERA, 1993–2009
The Age of Uncertainty 119

Part II. The Ultimate Lineup Card

9. FIELDING A TEAM OF GREAT HITTERS 141

10. THE TABLE SETTERS 182

11. THE TABLE CLEARERS 202

Part III. Hot Stove League Favorites Revisited

12. LEFT ON BASE 213

13. WHATEVER HAPPENED TO THE TRIPLE CROWN? 220

POSTGAME REPORT 228

Appendix: Using the BPPA Formula in Fantasy Baseball Leagues 233
Notes 235
Index 239

Preface

There are many different kinds of baseball books, but most of them can be broadly classified as anecdotal, biographical, historical, or statistical. This book is a combination of those elements—baseball statistics set in a historical framework supported by anecdotal and biographical data. All baseball books are concerned to some degree with statistics, since from its very beginning baseball has been heavily dependent on numbers. Using numbers is an indispensable means by which what takes place on the field is recorded and passed on to fans outside the ballpark and to fans in the future. That's the way it has always been and the way it will always be. There may be debates about what the numbers mean and which ones are better, but as long as there is baseball there will be numbers. The very permanence of the box score is perhaps the supreme testimony to that reality.

I have friends who tell me they are baseball fans but not numbers people. Yet, when we discuss baseball, numbers always seem to be involved. What my friends actually mean is that they are comfortable with their own baseball numbers and uncomfortable with any other baseball numbers. They seek to support their positions with their numbers. The other numbers, they say, are for statistics buffs, not true baseball fans. The real difference between me and my friends is not between baseball people and numbers people but between one kind of baseball numbers people and another kind. All baseball fans are, in one way or another, numbers people because baseball and numbers are inextricably intertwined.

You do not have to be a statistics buff in order to understand this book. You will encounter numbers, for it is the performance of baseball players that we are discussing and how else can you compare performance than through the use of numbers? Baseball fans can appreciate the personalities of baseball players and the finer points of the game as a craft without getting involved in statistics. But can they make meaningful comparisons between players and interpret what takes place on the field without statistics?

I have dedicated this book to my wife, Ruth Hager Taylor. She has enabled me to pursue this book at a time in life in which we had expected to be off somewhere smelling the roses. Instead, she has become the catalyst in this project—listener, typist, and emissary.

The other person whose help has been indispensible is my friend John Strunk, who guided me in using the computer as a more effective tool, reformatted my ragged draft into book form, and finalized the book for submission to the publisher. These tasks required enormous amounts of time, and I am very grateful for his help.

I am also very grateful for the invaluable help of my publisher, the Johns Hopkins University Press. Editor-in-Chief Trevor Lipscombe and Assistant Editor Greg Nicholl provided clear guidance and timely answers to questions. The Press staff have been both professional and accommodating. I appreciate the assistance of Kathy Alexander, Brendan Coyne, Julie McCarthy, Claire McCabe Tamberino, and Karen Willmes. I am especially indebted to Jeremy Horsefield, the copy editor, and Anne Whitmore, the production editor, with whom I have spent many hours reworking the manuscript. Everyone at the Press made the efforts of this first-time author an insightful and pleasurable experience.

My son, David, played the role of devil's advocate, and my daughter, Diana, helped in the search for a publisher. Derek Oldham helped on computer techniques. Dr. Wm. James Moore went out of his way to help solve computer problems. Wladimir Andreff reviewed part of the manuscript, and Lee Williams read the manuscript for consistency. I thank my entire family for supporting me in this project.

Abbreviations

AVG	batting average
BB	walks (bases on balls)
BPPA	bases per plate appearance
DBE	Dead Ball Era
DBI	Dead Ball Interval
ECC	Era of Constant Change
ERA	earned run average
ERPG	earned runs per game
ERS	expected runs scored
HP	hit by pitch
HR%	home run percentage
LBC	Live Ball Continued
LBE	Live Ball Era
LBEE	Live Ball Enhanced Era
LBI	Live Ball Interval
LBR	Live Ball Revived
LSLR	least squares linear regression model
LWTS	linear weights
OBP	on-base percentage
OPS	on-base plus slugging
PRG	potential runs per game
RBI	runs batted in
RBI%	runs batted in percentage

RC   runs created
SLG   slugging average
SST   Strength, Skill, and Timeliness Award
TA   total average
TBBA   total batter bases average

The Runmakers

Pregame Analysis

> When you can measure what you are speaking about, and express it in numbers, you know something about it; but when you cannot measure it, when you cannot express it in numbers, your knowledge is of a meager and unsatisfactory kind: it may be the beginning of knowledge, but you have scarcely, in your thoughts, advanced to the stage of science.
>
> **Lord Kelvin of England,** *Popular Lectures and Addresses* (1891–1894)

If Lord Kelvin were reincarnated as a contemporary baseball fan, he would be pleased. Baseball kept detailed records from its very beginning and expanded the amount of information recorded as the seasons wore on. Today, newspapers, magazines, yearbooks, encyclopedias, radio and television programs, and Web sites abound with baseball information. And this information is comprehensive—one can find the day-by-day statistics for every active player and the season-by-season statistics for every player ever involved in Major League Baseball. It is truly amazing. Baseball today is awash in a veritable sea of statistics.

The Measure of a Hitter
One of the great questions about baseball statistics is how to use them to measure and evaluate hitters. There are many ways to do this, and there is no consensus on which way is the best. Some baseball people prefer to use total numbers (hits, home runs, runs batted in, etc.), but averages (batting average, on-base percentage, slugging

average, etc.) correlate more closely with runs scored, and that is, after all, the objective of the game. If you measure careers by the total amount produced, you are favoring some players with long careers over other players with shorter but more successful careers. Thus, many baseball people prefer to measure and evaluate players based on their averages—that is, the rate at which players produced rather than the total amount they produced. In essence, it all comes down to a preference for quality over quantity.

The three most popular average measurements—batting average (AVG), on-base percentage (OBP), and slugging average (SLG)—have been around for a long time. They are well known and widely used throughout the baseball community. A fourth measurement has more recently gained some popularity: on-base plus slugging (OPS). There are also several more technical measurements, including Tom Boswell's total average (TA), Pete Palmer's linear weights (LWTS), Bill James's runs created (RC), and Jim Albert and Jay Bennett's regression model (LSLR). This book introduces a ninth measurement called bases per plate appearance (BPPA).

Most of these average measurements apply weights to various batting events. Only two of them—batting average and on-base percentage—are based solely on raw, unweighted player data. All baseball fans know that batting average is calculated by dividing a player's hits by his at bats. Henry Chadwick, the father of baseball (1824–1908), was an early advocate of batting average as the best measure of a hitter. When the National (1876) and American (1901) Leagues were formed, they decided that the player with the highest batting average would be their official batting champion, and, except for a very short time, that practice has continued to this day. Knowledgeable baseball fans know that on-base percentage is calculated by dividing the sum of a player's hits, walks, and hit by pitch by the sum of his at bats, walks, hit by pitch, and sacrifice flies. This measurement was developed by Branch Rickey's statistician, Alan Roth, in the early 1950s, and the major leagues adopted it as an official statistic in 1984.

All of the other average measurements apply weights to batting events to arrive at estimates of total bases. Ferdinand Lane, editor of *Baseball Magazine*, rejected batting average in favor of power baseball and in 1915 developed a system in which weights were applied to walks, singles, doubles, triples, and home runs, a precursor to slugging average and other subsequent measures of offensive value. Most baseball fans know that slugging average is total bases (singles times one, doubles times two, triples times three, and home runs times four) divided by at bats. The National League began to record slugging average in 1923, but the American League did not do so until 1946.

From 1977 to 1988, Bill James published the annual *Baseball Abstract* in which he introduced a new formula for assessing batter performance, which he called runs created. The core of that formula was hits plus walks times total bases divided by at bats plus walks. He later developed a series of 24 similar but slightly different formulas called historical data groups, based on the availability of data for the years in which each of the formulas was to be used. Once he had calculated runs created, James divided it by another formula to arrive at runs created per 27 outs (RC/27), which he defined as "the number of runs per game that a team made up of nine of the same player would score."[1]

Thomas Boswell, a sports columnist for the *Washington Post*, introduced his total average (TA) formula in 1981. Total average is essentially total bases divided by total outs. Barry Codell had introduced a similar formula in 1979. It has the elements in Boswell's formula plus sacrifice hits and sacrifice flies, which he included in both parts of his formula. Boswell's formula was easier to use and became better known in the baseball community.

In 1984, statistician Pete Palmer and writer John Thorn published *The Hidden Game of Baseball*, which introduced Palmer's linear weights (LWTS) model, which is essentially total weighted bases minus 25% of batter outs (AB-H) and 60% of times caught stealing. The result is an estimate of the number of runs produced, which can be converted to an average by dividing it by the number of games played. Palmer also developed the on-base plus slugging (OPS) model, which correlated closely with linear weights and could be calculated more easily. The *New York Times* helped popularize OPS by publishing OPS statistics regularly in the mid-1980s.

In 2001, Jim Albert and Jay Bennett changed the weights in Boswell's total average formula and arrived at a new formula that they named least squares linear regression (LSLR) after the statistical technique they used to arrive at the weights applied to the batting events. This measurement is essentially total weighted bases divided by the number of games. The events used to calculate total bases in the LSLR formula are the same as those in the total average and linear weights formulas. The basic difference between the three formulas is the value of the weights applied to the batting events and the opportunity factors used to calculate the averages.

The BPPA measure uses a different perspective. Runs are scored in a step-by-step series of events: getting on base, advancing from one base to another, and finally scoring. Sometimes runs are scored all at once with a home run. Most of the time, however, scoring runs requires teamwork—different players share in getting on base, advancing runners from base to base, and driving runners home. Thus, one can add the number of bases players accumulate in getting on base, advancing base runners, and driving in runs and divide the total by the number of plate appearances to get the number of bases per plate appearance. The BPPA should correlate closely with the number of runs scored because both are measures of productivity in the interrelated process of accumulating bases and scoring runs. The higher the rate of producing bases, the higher the rate of scoring runs. Bases add up to runs—take care of the bases and the runs will take care of themselves.

Bases per Plate Appearance Measure

Before proceeding to an evaluation of the average measurements, we need to digress in order to consider the construction of the BPPA model. The total bases calculation in this model has three parts: (1) Bases earned by batters for themselves are calculated by applying a weight of one to walks (BB), hit by pitch (HP), and singles; two to doubles; three to triples; and four to home runs. (2) Bases earned for advancing runners include one base for advancing a runner from first base to second base or from second base to third base, and two for advancing a runner from first base to third base. (3) Bases earned for runs batted in from the bases include three bases for a runner scoring from first base, two bases for a runner scoring from second base, and one base for a runner scoring from third base.

In building the parts of the BPPA model dealing with runners advanced and runs batted in from the bases, four sets of assumptions were made. The first set of assumptions had to do with estimates of team plate appearances with runners on first base, second base, and third base. Two sets of plate-appearance assumptions were made—one set for all National League teams from 1876 to 2009 and American League teams from 1901 to 1972 (prior to adoption of the designated hitter rule) and another set for American League teams from 1973 to 2009.

The second set of assumptions had to do with the percentage of runners advancing from first base to third base versus advancing from first base to second base on singles and the percentage of runners scoring from first base versus advancing from first base to third base on doubles.

The third set of assumptions dealt with the number of batting events that were relevant to the advancing of runners and driving in runs from the bases. The initial list included 12 events—walks, hit by pitch, singles, doubles, triples, home runs, runs batted in minus home runs (RBI from the bases), stolen bases, caught stealing, sacrifice hits, sacrifice flies, and grounded into double plays.

The fourth set of assumptions had to do with the performance of teams. It was assumed that in any given year, teams performed uniformly—that is, the same in situations with runners occupying various bases as with no runners on base. Thus, over the course of a complete season it was assumed that there were no clutch-hitting teams.

The four sets of assumptions were combined to develop team factors for runners advanced and runs batted in from the bases. The bases derived from applying these factors were added to the bases batters earned for themselves to arrive at total bases per team, which were then divided by plate appearances to obtain total bases per plate appearance. The team BPPA estimates were then paired with actual team runs per game for each team and year and tested statistically, resulting in high correlation coefficients. Runs per game was selected instead of total runs because in many years some teams played more games than other teams and for that reason would tend to score more runs than other teams and distort the data.

So far so good, but the process was continued in order to improve the results as much as possible. The first, second, and third sets of assumptions were then incrementally changed in a long series of trial-and-error tests in order to find which combination of assumptions correlated best with runs per game. The fourth set of assumptions remained fixed. The trial-and-error process was continued until the BPPA estimates and runs scored correlation coefficients were maximized.

At this point, it is important to remember that we have been talking about measuring teams, not players! Before considering the measurement of players, however, we need to return to our narrative and consider how the various measurements compare and test against one another on a team basis.

Scouting How Teams Measure Up

The various measurements rate and rank teams differently, sometimes markedly so. The number 1 team for one measurement might be the number 10 team for another measurement, or the number 5 team for one measurement might be the number 16 team for another measurement, etc. The question then arises, how do we settle these differences? How do we decide which measurement to use?

I was pondering the answer to this question one evening when I suddenly remembered that it had already been answered. Several years ago, the commissioner of baseball appointed a blue ribbon committee to answer this question. It was a committee of eight experts composed of leading figures in the baseball world. After the various measurements were fully explained, discussed, and debated, the committee decided to take a vote. It was discovered that each of the committee members was an advocate for a different measurement and none of them would admit that their measurement was inferior to any of the others. They were so locked in to their respective measurements that they could not compromise. The committee was deadlocked with one vote for each of eight different measurements.

It was pointed out that baseball logic, backed by statistical evidence, demonstrated that the weighted measures were better than the unweighted measures, and the advanced weighted measures were better than the basic weighted measures. An attempt was made to form alliances within the committee. The six weighted measurement advocates easily outvoted the two unweighted measurement advocates, but they couldn't agree among themselves. The three advanced weighted measurement advocates disagreed with the three basic weighted measurement advocates. The committee members were still arguing with each other when I woke up. Yes, it had been merely a dream—no such committee was ever appointed.

As I later thought more about it, I concluded that the real baseball world in effect operates much like the blue ribbon committee operated in the dream world and the result in each world is essentially the same—a stalemate. The advocates of each measure advance their theories, but no theory is recognized as the best; indeed, there is hardly any differentiation between them: in both worlds they are all essentially the same. Major League Baseball recognizes on-base percentage and slugging average as official statistics but awards its annual batting titles to the players with the highest batting averages. The baseball press recognizes those three statistics and on-base plus slugging but does not deal with the issue of accuracy. There is no discussion or debate about the strengths and weaknesses of each. Everyone goes his own way.

There is a more logical way to approach this problem. Batting average and on-base percentage may be official major league statistics, but they do not reflect the official rules by which the game is played. On the field, batters are awarded one base for a single, two bases for a double, three bases for a triple, and four bases for a home run, but in these two official statistics all extra-base hits count only one, the same as a single. Since the objective of a baseball team is to score more runs than their opponent, the value of a hitter should be based on his contribution to the scoring of runs. Thus, the evaluation of a hitter is analogous to the valuation of money. All hits, like all coins, are not of equal value. To determine the total value of your coins, you add up the monetary value of each—one for a penny, five for a nickel, ten for a dime, etc. When you determine the value of a hit, you should consider the value of different kinds of hits. You wouldn't think of treating all coins as equal. Why then should you treat all hits as equal (as in batting average) or all times on base as equal (as in on-base percentage) when you measure the value of a player? Batting average has the additional liability of not counting for the impact of walks and hit by pitch.

The rest of the average measures apply weights to batting events, but both the value of the weights and the batting events themselves differ. Slugging average is the least credible weighted model because it has the fewest number of events in its formula. It excludes walks, hit by pitch, and stolen bases, events that have a definite impact on the scoring of runs. OPS is a dubious measurement because it incorporates the shortcomings of both on-base percentage (it does not weight batting events) and slugging average (it does not include events included in other weighted models). Total average is the next least credible model because it applies the same weight of one to singles, walks, hit by pitch, and stolen bases. The last three of these events are not equal to a single in their potential for scoring runs. With the bases empty, a walk may be as good as a single, but with runners on base, a single is much more potent because singles can advance runners two bases and walks advance runners only one base and then only when there is a runner on first base.

The advanced weighted measures—LWTS, RC/27, and LSLR—apply unique weights to the batting events. The linear regression and linear weights models are very close to each other in the weights they apply. But how can a baseball fan tell whether a home run should have a weight of 1.40 or 1.50 or whether a walk should have a weight of .33 or .36? Baseball itself provides no help in answering these kinds of questions. We have to go to the world of statistics to find the answer.

Jim Albert and Jay Bennett conducted a series of statistical tests to find out which of various measurements was the best predictor of team runs per game.[2] All the average measurements used in this book, except bases per plate appearance, were included in their tests. They found that their LSLR model was best, followed closely by runs created and linear weights, then by total average and on-base plus slugging, and finally by slugging average, on-base percentage, and batting average. The last three were clearly the least accurate.

After the BPPA formula was developed, it was paired with team runs per game and tested statistically for the 109 years from 1901 to 2009. The 25-year period from 1876 to 1900 was not tested because all of the information for those years was not available. Besides, the rules of the game were significantly different during the early years of baseball. The BPPA model tested better than the LSLR model in 104 of the 109 years. The average BPPA/runs per game correlation coefficient was .9735 versus .9438 for the LSLR/runs per game pairing. The average coefficient of determination for the BPPA/runs per game pairing was .9477 versus .8908 for the LSLR/runs per game pairing. The standard error of the estimate for the BPPA/runs per game pairing was better than the LSLR/runs per game pairing in 101 of the 109 years. For the 1954 to 1999 period, the BPPA model predicted two-thirds of the team runs within .102 runs per game, which was significantly closer than the .142 runs per game reported by Albert and Bennett for their model for the same period of time.

If the BPPA model is more accurate than the regression model and the regression model is more accurate than all of the other models, the BPPA model must also be more accurate than all of the other models for predicting team runs per game. It is logical that the BPPA model is the most accurate for predicting team runs per game because it alone includes specific factors for advancing runners and driving in runs, which are essential steps in the process of scoring runs. The other weighted models merely assume that the more bases teams earn for themselves, the more bases they earn for advancing runners and for driving in runs, but this is not necessarily so.

Applying the Bases per Plate Appearance Measure to Players

Up to this point we have been talking about applying the BPPA formula to teams. This formula cannot ipso facto be applied to players because all players do not have an equal opportunity to advance runners and drive in runs. Two conditions work against that—the impact of the batting order and the designated hitter rule. The scoring of most runs requires teamwork. A batter gets on base, another batter advances him along, and a third batter drives him home. Only in the case of a solo home run is just one batter involved in scoring a run. This fact of baseball life, by the way, is perhaps the biggest reason why fantasy baseball leagues remain fantasies instead of realistic replications of reality. The BPPA formula offers a solution to this problem (see the appendix).

Some players are good at getting on base, others at advancing runners on the bases, and still others at driving in runs. Batting orders are constructed with this division of labor in mind. Players who are good at getting on base are picked to be leadoff batters. Players who are good at advancing runners are picked to be number 2 batters. Players who hit the long ball and have RBI potential are concentrated in the heart of the order (numbers 3, 4, and 5). Thus, leadoff batters and number 2 batters are sometimes referred to as table setters and number 3, 4, and 5 batters as table clearers. The latter are especially critical and the former, therefore, often neglected. This is reflected in the fact that only 16 of the more than 200 position players (non-pitchers) elected to the Baseball Hall of Fame were leadoff batters.

A second reality that works against equal player opportunity is the designated hitter rule. American League players during the designated hitter era have more opportunities to advance runners and drive in runs. This advantage applies not just to number 3, 4, and 5 hitters but throughout the lineup. Designed hitter era leadoff batters and number 2 batters also have more opportunities than their non–designated hitter league counterparts.

It was reasoned that the conditions working against the equal opportunity of players to advance runners on the bases and drive in runs from the bases could be neutralized by separating players into three distinct groups—those batting first, those batting second, and those batting third, fourth, and fifth in the batting order. The sixth to ninth positions in the batting order were not addressed. We are dealing here with the leading hitters of a historical era or for a position played, and players who perform at that level would not have been batting that low in the batting order.

Each of these three groups of batters were then given runner-advancement and runs-batted-in-from-the-bases factors extrapolated from the team factors—one set for those who played in the National League and the American League from 1901 to 1972 and another set for those who played in the American League from 1973 to 2009. The factors for the third, fourth, and fifth batters were higher than the overall team factors because they were preceded in the batting order by leadoff batters and number 2 batters who were put in those positions largely because of their ability to get on base. Thus, third, fourth, and fifth batters had more plate appearances with runners on base.

The factors for leadoff batters were lower than the overall team factors because in their first plate appearance in every game (perhaps 20% of a season's total plate appearances) there was never a runner on base and in their subsequent plate appearances they batted after the ninth batter—either the pitcher or some other weak-hitting player. The leadoff batters thus had many fewer plate appearances with runners on base. The factors for number 2 batters were set in between those for leadoff batters and third, fourth, and fifth batters—a little closer to the former than to the latter.

A logical question arises: how accurate are these player factors? There are two parts to the answer to this question. First, it is acknowledged that there is no proof that the player BPPA measures are better than the other measures—nor, however, is there any proof that any other measure is better than the player BPPA measures. Second, it can be argued that the BPPA player measures should be the best because (a) they were extrapolated from the BPPA team measures, which had tested best, and (b) the margin by which the team

BPPA measures exceeded the second-best LSLR measure should be sufficient to offset any loss in the extrapolation process.

The differences between the team and player measures are important, but they are not that great. If the team factors were applied to the third, fourth, and fifth American League batters instead of the player factors, the BPPA ratings would be only 0.6% lower (e.g., .895 vs. .900). If the team factors were applied to the same group of National League batters, the BPPA ratings would be 1.3% lower (e.g., .888 vs. .900). Surely the margin by which the BPPA team measures—the base from which the player measures were derived—exceeded the second-best LSLR measure would more than offset these differences.

The application of the BPPA team factors to players would not, at any rate, change the order of their ranking—their BPPA ratings would change, but they would be reduced proportionately. It is important to point out that the potential runs per game (PRG) and LSLR player rankings for each historical era and for each position in the field are actually quite close to each other. For the historical eras, the two measures agree exactly on the top 10 rankings of an average of 2.75 players, and on average 5.13 players are within two rankings of each other. For the player positions (excluding designated hitters), the two measures agree exactly on the top 10 rankings of an average of 2.25 players, and on average 5.6 players are within two rankings of each other.

It is also important to keep in mind that it is not even certain that the player BPPA measures are any less accurate than the team BPPA measures. They could actually be more accurate, but there is no way to test their accuracy by pairing BPPA player estimates with player runs. Players are not solely responsible for most runs because they are scored by the efforts of two or more players.

In conclusion, there is no proof that the BPPA measures, or any other measures, are the best for ranking players. There is evidence to suggest, however, that the BPPA measures may be the best. They are summarized in table I.1, which is followed by tables that constitute examples of how the factors are applied to players in various categories (tables I.2–I.7).

As you proceed through this book, you will find very significant differences in the way different measures rank players. If you want to identify the best-hitting player for a historical era, for a position in the field, or for a position in the batting order, you have to choose one of those measures or a combination of those measures or come up with one of your own. That choice should not be an arbitrary one—based, for example, on your favorite statistic or the statistic that ranks highest your favorite player or the player who played on your favorite team. There should be a rationale for the measure you choose, and, in the interest of consistency, the measure you choose to rank players from one historical era or position should also be the one you choose to rank players from the other historical eras and positions. There is no justification for selecting one measure for one historical era or position and another measure for another historical era or position.

Rating and ranking people in different walks of life is a favorite activity of Americans, whether it be presidents, Hollywood actors and actresses, sports figures, etc. It is both fun and a challenge to participate in this game. You are invited to consider the rankings presented in this book and to second-guess them when you feel you have a point. Don't automatically accept what you read on the writ-

TABLE I.1 BPPA Factors

	National League (1876–Present) and American League (1901–1972)					American League (1973–Present)				
	BB+HP	S	D	T	HR	BB+HP	S	D	T	HR
Batters	1.00	1.00	2.00	3.00	4.00	1.00	1.00	2.00	3.00	4.00
					Runners advanced					
Leadoff batters	.273	.318	.219	N/A	N/A	.320	.380	.271	N/A	N/A
Second batters	.351	.429	.303	N/A	N/A	.380	.470	.336	N/A	N/A
3, 4 & 5 batters	.390	.484	.344	N/A	N/A	.409	.514	.368	N/A	N/A
					RBI from the bases					
Leadoff batters	.021	1.595	2.084	2.331	2.331	.021	1.600	2.150	2.400	2.400
Second batters	.021	1.602	2.133	2.397	2.397	.021	1.604	2.170	2.420	2.420
3, 4 & 5 batters	.021	1.605	2.157	2.430	2.430	.021	1.610	2.200	2.510	2.510

TABLE I.2 BPPA Formula Applied to Babe Ruth (Middle-of-the-Order Batter, American League before 1973)

	BB/HP	S	D	T	HR	Bases	PA	BPPA
Events	2,105	1,517	506	136	714			
Batter bases	2,105	1,517	1,012	408	2,856	7,898		
Runners advanced bases	821	734	174	N/A	N/A	1,729		
RBI bases						2,940[1]		
TOTAL						12,567	10,617	1.184

[1] 1,499 RBI from the bases (2,213 RBI – 714 HR) × 1.961 bases per RBI = 2,940 RBI bases. The 1.961 bases per RBI is derived by multiplying each event by the appropriate factor from table I.1 to get total bases, which is then divided by the number of hits [(44 + 2,435 + 1,091 + 330 + 1,735)/2,873 = 1.961].

TABLE I.3 BPPA Formula Applied to Albert Belle (Middle-of-the-Order Batter, American League after 1972)

	BB/HP	S	D	T	HR	Bases	PA	BPPA
Events	738	935	389	21	381			
Batter bases	738	935	778	63	1,524	4,038		
Runners advanced bases	302	481	143	N/A	N/A	926		
RBI bases						1,683[1]		
TOTAL						6,647	6,673	.996

[1] 858 RBI from the bases (1,239 RBI – 381 HR) × 1.961 bases per RBI = 1,683 RBI bases. The 1.961 bases per RBI is derived by multiplying each event by the appropriate factor from table I.1 to get total bases, which is then divided by the number of hits [(15 + 1,505 + 856 + 53 + 956)/1,726 = 1.961].

TABLE I.4 BPPA Formula Applied to Augie Galan (Leadoff Batter, American League before 1973)

	BB/HP	S	D	T	HR	Bases	PA	BPPA
Events	1,004	1,196	336	74	100			
Batter bases	1,004	1,196	672	222	400	3,494		
Runners advanced bases	274	380	74	N/A	N/A	728		
RBI bases						1,298[1]		
TOTAL						5,520	7,003	.788

[1] 730 RBI from the bases (830 RBI – 100 HR) × 1.778 bases per RBI = 1,298 RBI bases. The 1.778 bases per RBI is derived by multiplying each event by the appropriate factor from table I.1 to get total bases, which is then divided by the number of hits [(21 + 1,908 + 700 + 172 + 233)/1,706 = 1.778].

TABLE I.5 BPPA Formula Applied to Wade Boggs (Leadoff Batter, American League after 1972)

	BB/HP	S	D	T	HR	Bases	PA	BPPA
Events	1,435	2,253	578	61	118			
Batter bases	1,435	2,253	1,156	183	472	5,499		
Runners advanced bases	459	856	157	N/A	N/A	1,472		
RBI bases						1,580[1]		
TOTAL						8,551	10,740	.796

[1] 896 RBI from the bases (1,014 RBI − 118 HR) × 1.763 bases per RBI = 1,580 RBI bases. The 1.763 bases per RBI is derived by multiplying each event by the appropriate factor from table I.1 to get total bases, which is then divided by the number of hits [(30 + 3,605 + 1,243 + 146 + 283)/3,010 = 1.763].

TABLE I.6 BPPA Formula Applied to Mark Koenig (Second Batter, American League before 1973)

	BB/HP	S	D	T	HR	Bases	PA	BPPA
Events	233	918	195	49	28			
Batter bases	233	918	390	147	112	1,800		
Runners advanced bases	82	394	59	N/A	N/A	535		
RBI bases						724[1]		
TOTAL						3,059	4,603	.665

[1] 415 RBI from the bases (443 RBI − 28 HR) × 1.745 bases per RBI = 724 RBI bases. The 1.745 bases per RBI is derived by multiplying each event by the appropriate factor from table I.1 to get total bases, which is then divided by the number of hits [(5 + 1,471 + 416 + 117 + 67)/1,190 = 1.745].

TABLE I.7 BPPA Formula Applied to Derek Jeter (Second Batter, American League after 1972)

	BB/HP	S	D	T	HR	Bases	PA	BPPA
Events	1,028	2,027	438	58	224			
Batter bases	1,028	2,027	876	174	896	5,001		
Runners advanced bases	391	953	147	N/A	N/A	1,491		
RBI bases						1,507[1]		
TOTAL						7,999	9,809	.815

[1] 844 RBI from the bases (1,068 RBI − 224 HR) × 1.786 bases per RBI = 1,507 RBI bases. The 1.786 bases per RBI is derived by multiplying each event by the appropriate factor from table I.1 to get total bases, which is then divided by the number of hits [(22 + 3,251 + 950 + 140 + 542)/2,747 = 1.786].

ten page—it's important to be informed, but it's also important to be independent.

In applying the BPPA model to individual players, I have converted the unit of accumulating bases per plate appearance to potential runs per game (PRG). This makes it much easier to appreciate a player's contribution to his team. If a player, for example, averages one base per plate appearance, it means that he has the potential for producing 1.25 runs per game (one base per plate appearance times an average of five plate appearances per game divided by four bases per run equals 1.25 potential runs per game). In the 2009 season, major league teams averaged 4.63 runs per game. Thus, a player with a BPPA rating of 1.0 had the potential for producing 1.25 runs per game, the equivalent of 27% (1.25 divided by 4.63) of an average team's runs. Seen in this context the impact of a hitter's performance on his team is much clearer than if we relied on BPPA alone. The conversion process does not affect the relative rating and ranking of the players.[3]

The lists of leading hitters in this book include only players with 4,000 or more career plate appearances. A lower plate-appearance minimum would have resulted in longer, unmanageable lists of players. A higher plate-appearance minimum would have resulted in the exclusion of many players who played long enough to make their mark and deserve to be considered.

The Game Plan

This book identifies the leading hitters of each baseball era, for all time, and by position, according to each of the models discussed above. Players are rated higher or lower depending on the model. Substantial differences between the models are addressed and reasons given why some of the models are better than others. The background and major accomplishments of the leading hitters are summarized in thumbnail sketches.

Part I measures and evaluates hitters by historical era—one chapter for each of eight historical eras. Most baseball writers pick milestone events—such as the beginning or ending of World War II, the expansion of the number of teams, or the adoption of the designated hitter rule—as markers for the ending of one era and the beginning of another. This book, instead, bases the identification of eras on performance, on what matters most in baseball, the scoring of runs—using the number of earned runs per game (ERPG) as the criterion.[4]

Table I.8 summarizes ERPG and other measures by historical era, not in the usual chronological order, but in descending order from the eras with the highest ERPG to the eras with the lowest ERPG. The first two historical eras—the Era of Constant Change and the Live Ball Interval—are excluded from this table because the conditions under which baseball was played then were so different from those in the subsequent eras that meaningful comparisons are doubtful at best. The table reflects an amazing consistency between the order of the measures and ERPG by era. In only OPS and PRG is the order of the eras different from that in the other measures, and then only with regard to the Live Ball Continued and the Live Ball Revived eras, in which the order of the two eras is the same (PRG) or reversed but only slightly different (OPS). The data in the table thus provide strong support for dividing baseball history into the historical eras chosen.

Chapter 1 looks back to the early years of Major League Baseball (1876–1892), a time when hitters were constantly challenged by changes in the rules, teams, and leagues. Chapter 2 moves on to

TABLE I.8 Earned Runs per Game and Weighted Measures by Era

Era	Weighted Measures						
	ERPG	PRG	LSLR	LWTS	TA	OPS	SLG
Live Ball Enhanced (1993–2009)	4.40	.960	7.64	+1.772	.719	761	.423
Live Ball Era (1921–1941)	4.13	.943	7.35	+1.126	.662	742	.398
Live Ball Revived (1977–1992)	3.84	.896	7.14	+.572	.646	711	.387
Live Ball Continued (1942–1962)	3.83	.896	7.10	+.511	.639	712	.382
Dead Ball Interval (1963–1976)	3.52	.851	6.78	−.183	.607	697	.368
Dead Ball Era (1901–1920)	2.88	.809	6.29	−1.494	.558	651	.334

consider the hitters of the next era (1893–1900), a short period of relative stability in which the ball was "alive" and batters prospered. In chapter 3, I examine the subsequent period of retrenchment (1901–1920), when batters played at a great disadvantage because the ball was "really dead," while in chapter 4, I highlight the statistics of those who played in the interwar period (1921–1941), when the ball came alive and the game was radically changed forever. Chapter 5 moves on to measure and evaluate hitters during and after World War II (1942–1962), when the level of hitting declined a little but remained at a high level. In chapter 6, I appraise those hitters who had to play at a time when the ball once again was dead (1963–1976). Chapter 7 looks at the accomplishments of those who played during the subsequent revival of the live ball (1977–1992). In chapter 8, I consider the statistics of the current era (1993–2009), a time when the legitimacy of an unprecedented increase in hitting is being widely questioned in the baseball community and beyond.

The terms *live ball* and *dead ball* are, of course, metaphors. Essentially they are used to distinguish batters' eras from pitchers' eras. The ball itself has not been literally alive or dead, although the condition of the ball has been one of many factors involved. At times the ball has been made more lively through, for example, the replacement of a rubber center with a cork center, tighter winding of the yarn, or the more frequent replacement of old balls with new balls. But the ball itself has always been only one of many factors contributing to the rise and fall of the number of runs scored. Among the other factors are the distance from home plate to the pitcher's mound, the height of the pitcher's mound, and the size of the strike zone.

Part II measures and evaluates the best hitters of all time. Chapter 9 ranks the best number 3, 4, and 5 hitters ever at each position in the field. Chapter 10 ranks the best leadoff and number 2 batters of all time. Chapter 11 ranks the best number 3, 4, and 5 hitters ever regardless of position played in the field.

Part III addresses several special subjects relevant to the performance of hitters. Chapter 12 addresses some controversial issues, including clutch hitting, streakiness, the impact of baseball stadiums, and others said to influence batting performance. Chapter 13 discusses the once revered but now almost forgotten jewel of baseball, the Triple Crown, and suggests a new award that would continue to recognize the skills involved in Triple Crown hitting.

In the closing comments of the Postgame Report, the major findings of the book are summarized.

PART I

Every Era Has Its Greats

* 1 *

The Era of Constant Change, 1876–1892
THE AGE OF DAN BROUTHERS

First Baseman "Big Dan" Brouthers was the first great slugger in baseball. Not only was he a remarkable percentage hitter, leading his league in batting average five times, he also was a great home run hitter . . . six consecutive times he led the NL in slugging. . . . At age 36, he hit .347 with 39 doubles, 23 triples, and 128 RBIs, and he also had a career-high 38 stolen bases.

The Baseball Encyclopedia, pp. 131 and 132

Contrary to popular belief, the game of baseball was not invented by Abner Doubleday at Cooperstown, New York, one day back in 1839. The roots of baseball actually go back to the eighteenth-century English game of rounders.[1] In America, rounders changed into town ball and town ball changed into baseball. Baseball caught on so well that some wanted to make it a distinctively American sport, our national pastime. In 1845 Alexander Cartwright, founder of the Knickerbocker Club of New York, published the first set of modern baseball rules. In its early years baseball was strictly a recreational game played for fun by amateurs, but as it became more popular, teams began paying players and baseball became a game, a profession, and a business.

The first professional baseball league, the National Association of Professional Baseball Players, was established in 1871 but lasted only five years. The current National League, chartered as the National League of Professional Baseball Clubs, was founded in 1876. The National League had several early rivals, but by 1891 it remained the only major league.[2] The American League wasn't established until 1901.

Casual baseball fans may believe that Babe Ruth ushered in the live ball era and everything before him constituted the dead ball era. The first part of this belief is true, but the second is not. The entire period before Babe Ruth should not be labeled a dead ball era. From 1893 to 1900 the ball was alive and batters prevailed over pitchers. This live ball interval didn't last very long, but it was very real and very distinct from what came before and what came after. The period before 1893 was a dead ball era, but it was a period replete with so many changes that it can more appropriately be called the Era of Constant Change (ECC).

The hallmark of the early years of professional baseball was change, both organizational and rule-based, as baseball tried to consolidate itself both as a game and as a business. Organizationally, there were frequent changes in the number of competing leagues and the number of teams in those leagues. The number of teams in each major league is listed by year in table 1.1. The total number of major league teams in any one year varied considerably, ranging from eight in 1877 and 1881 to 34 in 1884. From 1876 to 1892, 22 different cities had a team in the National League at least one year; only two cities—Boston and Chicago—had a team in the league every year, and those two cities won 11 of the 17 pennants. From 1882 to 1891, the American Association had 19 different cities with a team in the league at least one year; only four cities had a team in the league every year, and those teams won 6 of the 10 pennants.

Frequent changes in the number of teams brought frequent changes in team rosters. Players changed teams often in the early years of baseball. Only two of the leading hitters, Cap Anson and Bid McPhee, played on only one team during their entire careers. Only one other leading hitter, Ned Williamson, played more than 10 seasons for one team, and the average leading hitter played on 4.5 different teams during his career. One leading hitter, Tom Brown, played on 11 different teams. Sixteen of the 23 leading hitters listed in table 1.3 and/or table 1.4 (70%) left their teams to play in the Player's League in 1890.

TABLE 1.1 The Number of Teams in the Early Major Leagues

Year	National League	International Association[1]	American Association	Union Association	Players League	Total
1876	8	—	—	—	—	8
1877	6	13	—	—	—	19
1878	6	23	—	—	—	29
1879	6	19	—	—	—	25
1880	8	19	—	—	—	27
1881	8	—	—	—	—	8
1882	8	—	6	—	—	14
1883	8	—	8	—	—	16
1884	8	—	13	13	—	34
1885	8	—	8	—	—	16
1886	8	—	8	—	—	16
1887	8	—	8	—	—	16
1888	8	—	8	—	—	16
1889	8	—	8	—	—	16
1890	8	—	9	—	8	25
1891	8	—	9	—	—	17
1892	12	—	—	—	—	12

[1] The International Association was a group of teams that operated more like a loose association than a strict league. It "was never a serious threat to the National League . . . its membership was too diffuse and scattered, not sufficiently well-knit and balanced to make possible a tight, economical playing schedule or to prevent a strong fight against the [National] League." The Canadian teams left the International Association after two years, and the new National Association disbanded after the 1880 season. The number of teams in the International Association is the best that could be ascertained from the sparse information available. Harold Seymour, *Baseball: The Early Years* (New York: Oxford University Press, 1960), pp. 99 and 102.

The frequent changes in the leagues, teams, and players had an impact on the level of play, but the biggest impact on the game came from the frequent changes in the rules, particularly those regulating the competition between pitchers and batters. The many changes in baseball rules affecting pitchers and batters between 1876 and 1900 are listed in table 1.2. In 1876 the pitcher delivered the ball from a 6 foot by 6 foot box. He was allowed a short run before pitching to a 12-inch-square home plate located only 45 feet away, and his hand had to pass below the level of his hips. The batter could call for a high or low pitch, and nine balls were required for a walk. Substitutions were allowed, but only in case of an injury or with the consent of the opposing team. The game is said to have resembled fast-pitch softball more than baseball.

The rules remained unchanged for three years, but beginning in 1880 there was at least one pitcher-batter rule change in each of 10 consecutive seasons. These changes moved the game away from its old softball look toward a more modern baseball look, essentially completed by 1893. That season the pitcher could deliver the ball with an underhand, sidearm, or overhand motion, from a 12 inch by 4 inch rubber slab, with no run before the pitch, to a home plate still 12 inches square but located 60 feet 6 inches away. The batter had to use a completely rounded bat, could no longer call for a high or low pitch, and it took four balls for him to reach first base on a walk. Free substitutions were allowed, but there could be no reentry of players removed from the game. Pitcher-batter rules were not settled but had been changed enough to achieve stability and bring an end to one era and a beginning to another.

Most of the rule changes favored batters over pitchers. It is easier to appreciate the impact of this by looking at earned run averages than runs per game because the high rate of errors at that time resulted in a significant number of unearned runs. The combined earned run average (ERA) of National League pitchers increased from 2.46 in 1876–1878 to 3.38 in 1890–1892. Batters benefited most from the gradual decrease in the number of balls required for a walk from nine to four. The rate of walks (per plate appearance) increased from 2.1% (1876–1878) to 8.9% (1890–1892). Although the early dominance of the pitcher was lessened during the Era of Constant Change, the era as a whole remained a pitchers' era. The ERA in the Era of Constant Change was only a little higher than the ERA in the Dead Ball Era (1901–1920) and was actually lower than the ERA in the Dead Ball Interval (1963–1976).

The constant organizational and rule-based changes affected the players. They made it very difficult to play well season after season since players were constantly adjusting to new leagues, new teams, and new rules. In an era when ERAs were increasing, it was espe-

TABLE I.2 Significant Changes in the Rules Affecting Pitchers and Batters: 1876 to 1900

Year	Pitcher's Box and Home Plate	Pitcher's Distance	Pitcher's Motion	Balls for a Walk	Batter
1876	PB: 6 × 6 ft. (with short run before pitch) HP: a 12 in. square	45 ft.	Hand must pass below the hip	9	May call for a high or low pitch. Must be thrown out at first base on a strikeout.
1880				8	Batter out on third strike held by catcher.
1881		50 ft.			
1882				7	
1883			Side arm pitch allowed		
1884			Shoulder-high pitch allowed	6	
1885					A portion of one side of the bat may be flat.
1886	PB: 4 × 7 ft.			7	
1887	PB: 4 × 5 1/2 ft.			5	No call for high/low pitch Extra strike on called 3rd strike.
1888					No extra strike on called 3rd strike.
1889				4	
1891					
1893	PB: 12 × 4 in. rubber slab (no run before pitch)	60 ft. 6 in.			Bat must be completely round, max. dia. 2½ in.
1895	PB: 24 × 6 in. rubber slab				Max. dia. of bat 2¾ in. Strike on a foul tip.
1900	HP: Five-sided shape 17 in. wide				

cially difficult for pitchers, and no pitcher was able to dominate the National League for very long. In 17 seasons 14 different pitchers led the National League in ERA, and only two of them did it more than once. On the other hand, it was also difficult for batters. Eleven different players led the National League in batting average, and only two of them did it more than once.

The top 10 hitters of the Era of Constant Change for each of the nine average measures of hitting performance are identified in table 1.3. If you look at the top 10 as a group, there is general agreement that eight particular players belong there. All measures agree on four players (Brouthers, Thompson, Connor, and O'Neill), all except one measure agree on three other players (Anson, Lyons, and Browning), and all except two measures agree on Bug Holliday.

If you look at the specific rankings of the players within the top 10, however, there is considerable disagreement. All measures do agree that Dan Brouthers is number 1, but there is much less agreement as you proceed. Sam Thompson is ranked number 2 by five measures, but three other players are ranked number 2 by four other measures. Thompson is ranked number 8 and number 9 by two other measures. Four other players—Anson, Connor, Browning, and Stovey— are ranked as high as number 3 by at least one measure, and all of them except Connor are ranked as low as number 10 by at least one measure. Thus, the lower the ranking, the more disagreement you encounter.

How do you reconcile all of this disagreement? In the Pregame Analysis it was acknowledged that there is no proof that the BPPA measure is better than any other measure for rating and ranking players—nor is there any proof of the opposite, that any other measure is better than the BPPA measure. I went on to say, however, that there is some evidence to suggest that the BPPA measure may be the best for rating and ranking players. That statement is not applicable to this particular era because the BPPA measure was not included in the statistical test that indicated that the BPPA measure was better than the LSLR measure. The baseball rules were significantly different in this era, there was a very high rate of errors, and the batting event information is not complete.

TABLE 1.3 The Top Ten Hitters in the ECC by Various Measures

Player	PRG	Advanced Weighted Measures			Basic Weighted Measures			Unweighted Measures	
		LSLR	RC/27	LWTS	TA	OPS	SLG	OBP	AVG
Dan Brouthers	1	1	1	1	1	1	1	1	1
Sam Thompson	2	2	8	2	5	2	2	9	3
Cap Anson	3	10	5	10	—	7	9	5	4
Roger Connor	4	5	4	4	3	3	3	4	6
Tip O'Neill	5	7	2	7	9	5	6	6	5
Denny Lyons	6	6	9	5	2	6	—	2	8
Henry Larkin	7	—	—	—	—	10	—	10	—
Oyster Burns	8	—	—	—	10	—	10	—	—
Bug Holliday	9	9	—	8	8	8	8	—	7
Buck Ewing	10	—	—	—	—	—	7	—	—
Pete Browning	(11)	4	3	3	4	4	4	3	2
King Kelly	(12)	—	7	—	—	—	—	—	10
Harry Stovey	(13)	3	6	6	6	9	5	—	—
Jim O'Rourke	(17)	—	—	—	—	—	—	—	9
Mike Griffin	(18)	—	—	9	7	—	—	7	—
George Gore	(20)	8	10	—	—	—	—	8	—

Before the BPPA measure was developed, Albert and Bennett determined that their LSLR measure was the best one available at that time. The best way to handle the disagreement between the various measures for this era is to compare the BPPA and LSLR rankings in table 1.3. They agree on (a) seven of the players included in the top 10, (b) the specific ranking of four players, and (c) the ranking of two other players within one or two ranks of each other.

The most significant differences between the two sets of rankings in table 1.3 concern Anson, Browning, and Stovey. Anson ranks higher in the PRG ranking because he ranked high in runs batted in from the bases, and Browning and Stovey rank lower in the PRG ranking because they ranked low in this category. The PRG model is the only one that uses runs batted from the bases as a specific factor. The most accurate ranking of batters for this era may lie somewhere between the two sets of rankings.

Table 1.4 summarizes the leading hitters of the Era of Constant Change, according to PRG, by the positions they played in the field. Three of the four highest-ranking hitters of the era were first basemen—Dan Brouthers (pronounced broothers), Cap Anson, and Roger Connor. Physically they fit the conventional image of first basemen. They were among the biggest men in baseball at that time. Dan Brouthers was 6 feet 2 inches tall, weighed 207 pounds, and was nicknamed "Big Dan." Cap Anson was 6 feet 2 inches tall and weighed 227 pounds. Roger Connor was 6 feet 3 inches tall and weighed 220 pounds. Brouthers and Connor batted and threw left-handed; Anson batted and threw right-handed.

The highest-ranking first baseman and the highest-ranking player overall was Dan Brouthers, with a PRG rating of 1.284. He was one of the most-traveled players of his time. His 19-year major league career was spent with 11 different teams. He played more than three years for only one team—the Buffalo Bisons of the National League. Dan Brouthers was a truly exceptionable player, hitting both for power and for a high percentage. He was Major League Baseball's first great slugger. Brouthers led his league in slugging average five times and had the highest slugging average for the era as a whole. He had the third most home runs (106) for the era, and several were so long that they were marked with flags at the ballparks. One of Brouthers's home runs at Capitol Park in Washington, D.C., was regarded as the longest hit in the early history of the game.

Brouthers led his league in batting average five times (four in the National League and one in the American Association). Although his career spanned all of the pitcher-batter rule changes listed in table 1.2, his first and last batting titles were 10 years apart. He was a consistently dominant hitter at a time when it was very difficult to hit well year after year. Brouthers's lifetime batting average of .342 ranks eighth all-time, sixth among left-handed batters, and first among first basemen. He was elected to the Baseball Hall of Fame in 1945 by the Old-Timers Committee.

The highest-ranking right fielder and the second-highest-ranking player overall was Sam Thompson, with a PRG rating of 1.283. Thompson played from 1885 to 1906: 10 years with the Philadelphia

TABLE 1.4 Leading Hitters of the ECC by Position[1]

Position	Leading Hitter	Second-Leading Hitter and Others in the Top 10 Overall
First base	Dan Brouthers — 1.284 (1)*	Cap Anson — 1.218 (3)*; Roger Connor — 1.183 (4)*; Henry Larkin — 1.155 (7)
Right field	Sam Thompson — 1.283 (2)*	Oyster Burns — 1.146 (8)
Left field	Tip O'Neill — 1.176 (5)	Harry Stovey — 1.088 (13)
Third base	Denny Lyons — 1.161 (6)	Billy Nash — 1.059 (14)
Center field	Bug Holliday — 1.121 (9)	Pete Browning — 1.109 (11)
Catcher	Buck Ewing — 1.111 (10)*	Jack Clements — 1.051 (15)
Second base	Hardy Richardson —1.044 (16)	Fred Pfeffer — .973 (27); Bid McPhee — .973 (27)*
Shortstop	Sam Wise — .999 (22)	Jack Rowe — .993 (25)

* Indicates that a player was elected to the Baseball Hall of Fame.
[1] The number after each player's name is his potential runs per game (PRG) rating, and the number in parentheses is his PRG ranking for the ECC.

Phillies, four years with the Detroit Wolverines (NL), and one year with the Detroit Tigers. He both batted and threw left-handed. At 6 feet 2 inches tall and 207 pounds, he was the same size as Brouthers, Anson, and Connor, but he was too fast and had too good a throwing arm for first base. He was one of the best outfielders of his day, possessing a strong arm, speed, and a fearless disdain for outfield fences, into which he often crashed.

Sam Thompson was an exceptional hitter, both a high-percentage hitter and a long ball hitter. He led his league in hits, runs batted in, and slugging average three times each, in doubles and home runs twice each, and in triples once. He ranks 10th all-time in runs batted in per at bat (.218). Some recognize Thompson as the best RBI man ever based on RBI per game. This statistic, however, favors those who played in the early days of baseball when the great number of errors gave batters more plate appearances per game and therefore more chances to drive in runs per game. The impact of the error factor is neutralized by relating runs batted in to at bats instead of to games.

Prior to Babe Ruth, Thompson's career total of 126 home runs was second only to Roger Connor's 138. He had a lifetime batting average of .331, fifth all-time for right fielders. Sam Thompson was elected to the Baseball Hall of Fame by the Veterans Committee in 1976.

The second-highest-ranking first baseman and the third-highest-ranking hitter overall was Cap Anson, with a PRG rating of 1.218. Anson played from 1872 to 1897, a 26-year career in which 22 years were spent with the Chicago White Stockings of the National League. Anson managed the White Stockings for 19 of those years, hence his nickname "Cap" for Captain. He is regarded as one of the greatest player-managers in the history of baseball. Unfortunately, he was also an early advocate of excluding black players from the major leagues, a tragic legacy that remained in effect for more than fifty years after Anson retired from the game.

Anson was one of the best-fielding first basemen in the league, but he was primarily known for his hitting, both percentage hitting and driving in runs. He led the National League in batting average four times and in runs batted in eight times. Anson did not strike out very often—he ranks 10th all-time in the number of at bats per strikeout (31). He has the fifth-highest all-time batting average for first basemen (.331), is ninth all-time for right-handed batters, and is 18th among all players in runs batted in per at bat (.207). Anson was elected to the Baseball Hall of Fame in 1939 by the Old-Timers Committee.

The third-highest-ranking first baseman and the fourth-highest-ranking hitter overall was Roger Connor, with a PRG rating of 1.183. Connor played from 1880 to 1897: 10 years with the New York Giants and eight years with four other teams. Connor was called "Dear Old Roger" because he was so popular with the New York fans.

Connor, like Anson, was one of the best-fielding first basemen in the league, but hitting was his main claim to fame. He has the distinction of hitting the first major league grand slam home run (1881). Connor hit more home runs (138) than any other player until 1921 when Babe Ruth broke his record. He led his league in every major hitting event, except runs, at least once. Connor is fifth all-time in triples (233). He also hit well for average, with a lifetime batting average of .317, or ninth place all-time for first basemen. Connor was elected to the Baseball Hall of Fame by the Veterans Committee in 1976.

The highest-ranking left fielder and the fifth-highest-ranking hitter overall was Tip O'Neill, with a PRG rating of 1.176. O'Neill played from 1883 to 1892: seven years with the St. Louis Cardinals (AA) and three years split among three other teams. He was more of a percentage hitter than long ball hitter. O'Neill's lifetime batting average of .326 was 11th all-time for right-handed batters, but he hit only 52 home runs. He had a phenomenal season in 1887, leading the league in nine major hitting events—runs, hits, doubles, triples, home runs, runs batted in, batting average, on-base percentage, and slugging average. In the entire history of Major League Baseball, this is the only time a player has led his league in doubles, triples, and home runs in the same season. It was also the only time that an American Association player ever won the Triple Crown. In 1886 he led the league in runs batted in and in 1888 in hits and average. These were the only years in which O'Neill led his league in any major batting event. His 1887 batting average of .435 is second only to Hugh Duffy, who hit .440 in 1894. O'Neill has never been elected to the Baseball Hall of Fame.

The highest-ranking third baseman and sixth-highest-ranking hitter overall was Denny Lyons, with a PRG rating of 1.161. Lyons played from 1885 to 1897: five years with the Philadelphia Phillies, four years with the Pittsburgh Pirates, and four years with four other teams. Denny Lyons was a good all-around player. He has the fifth-highest lifetime batting average (.310) and on-base percentage (.407) for a third baseman. Lyons was a good fielder, holding the single-season record for putouts for a third baseman (255) and second all-time for putouts per game at third base (1.55). He has never been elected to the Baseball Hall of Fame.

The fourth highest-ranking first baseman and seventh-highest-ranking hitter overall was Henry Larkin, with a PRG rating of 1.155. He played from 1884 to 1893: six years with the Philadelphia Athletics of the American Association, a year with the Cleveland Infants in the Players League, another year back with Philadelphia, and two years with the Washington Senators of the National League. He was a small right-handed-hitting first baseman who was a balanced hitter, not a consistent leader in slugging or batting average. He led his league in doubles twice and on-base percentage once. His best year was 1886, when he led the American Association in doubles and on-base percentage and had a career-high .319 batting average. Larkin has never been elected to the Baseball Hall of Fame.

The second-highest-ranking right fielder and the eighth-highest-ranking hitter overall was Oyster Burns, with a PRG rating of 1.146. Burns played from 1884 to 1895: five years with the Brooklyn Dodgers, four years with the Baltimore Orioles (AA), and two years split among three other teams. The origin of his nickname "Oyster" is unknown—his real name was Thomas. Burns was more of a slugger than percentage hitter. His lifetime slugging average was .446 and his batting average was .300. In 1887 he led his league in triples, and in 1890 he led his league in home runs and runs batted in, the only times he ever led his league in any major batting event. Oyster Burns has never been elected to the Baseball Hall of Fame.

The highest-ranking center fielder and the ninth-highest-ranking hitter overall was Bug Holliday, with a PRG rating of 1.121. Holliday played from 1889 to 1898: nine years with the Cincinnati Reds and one year with the Cincinnati Red Stockings (AA). The origin of his nickname "Bug" is unknown—his first name was James. Holliday was only 5 feet 10 inches tall and 151 pounds, but he led his league in home runs twice, the only batting event in which he ever led his league. He was a balanced hitter, ranking eighth in slugging average and seventh in batting average for the entire era. Bug Holliday has never been elected to the Baseball Hall of Fame.

The highest-ranking catcher and the 10th-ranked hitter overall was Buck Ewing, with a PRG rating of 1.111. Ewing played from 1880 to 1897: nine years with the New York Giants and nine years split among four other teams. Ewing's first name was William, but he was known as "Buck"—short for Buckingham, his middle name. Only twice did Ewing lead his league in a significant batting event (triples and home runs), but he was considered by many of his contemporaries as the best all-around player of his time. He was a good hitter, ranking seventh in the era for slugging and having a career batting average of .303. Ewing was also an excellent catcher with a strong arm, a savvy base runner, and an outstanding leader. He was elected to the Baseball Hall of Fame in 1939.

The shortstops and second basemen of this era were valued much more for their fielding ability than their hitting. This is reflected in the PRG rankings for the best-hitting second basemen. Hardy Richardson, Fred Pfeffer, and Bid McPhee are ranked 16th, 27th, and 27th, respectively. They were all good fielders, and in an era when so many errors were being made, it was especially important to have good fielders in the middle of the infield. Richardson was clearly the best hitter and McPhee second best. Richardson led his league in hits, home runs, and runs batted in once each, McPhee led his league in doubles and home runs once each, but Pfeffer never led his league in any major batting event. Bid McPhee was elected to the Baseball Hall of Fame by the Veterans Committee in 2000. Hardy Richardson and Fred Pfeffer have never been elected to the Baseball Hall of Fame.

The PRG ratings for the best-hitting shortstops also reflect the preference for fielding ability over hitting ability. Sam Wise and Jack Rowe ranked 22nd and 25th, respectively. Neither of them has been elected to the Baseball Hall of Fame.

Position players are elected to the Baseball Hall of Fame based on all of their accomplishments—at the plate, on the base paths, and in the field. Hitting, however, is probably the biggest factor in their election, and table 1.4 illustrates this point. Five of the 10 position players of this era elected to the Hall of Fame ranked in the top 10 in PRG—Dan Brouthers, Sam Thompson, Cap Anson, Roger Connor, and Buck Ewing. One player in table 1.4—Bid McPhee—was elected to the Hall of Fame for his fielding ability. McPhee ranked only 27th in hitting but was one of the best-fielding second basemen of his day. Three ECC players not in table 1.4 were elected to the Hall of Fame based on their fielding accomplishments. Jim O'Rourke was a good hitter, but his versatility in the field—he played more than 100 games at six different positions—was probably the primary factor in his election. John Ward was an average-hitting but good-fielding shortstop. Tommy McCarthy was only an average hitter, but his speed and strong arm made him an excellent defensive outfielder. King Kelly was also elected to the Hall of Fame based on his versatility. He was elected to the Hall of Fame as a catcher, although he actually played more games as an outfielder. He is treated here as a right fielder.

To summarize, the election of five ECC position players to the Hall of Fame was based on their hitting, the election of four was based on their fielding, and the election of one was based on his versatility.

All of the ECC players elected to the Baseball Hall of Fame spent all or most of their major league careers in the National League. This raises an interesting question. Why weren't American Association players elected to the Baseball Hall of Fame? The American Association was a legitimate major league. The National Agreement of 1883 recognized both the National League and the American Association as major leagues. Between 1884 and 1890 the two leagues played each other in a postseason series of games that was billed as the World Championship. The American Association won one and tied another of the seven series and won 22 of the 47 games (47%) played. Certainly the American Association teams were more competitive than the National League was against the American League in the World Series from 1999 to 2007. The National League lost six of the nine World Series and won only 16 of the 48 games played (33%). Surely one would not advocate excluding National League players in those years from the Baseball Hall of Fame. Perhaps some American Association players in the Era of Constant Change should have been elected to the Baseball Hall of Fame.

We are left with one unanswered question about the leading hitters of this era. How do they compare with the leading hitters of subsequent eras? Unfortunately, ECC players were participants in a hybrid sport so different from the game played in subsequent eras that there is no basis for comparing their statistics to those of the players who followed. The rules of the game during this era were very different from those in subsequent eras, and the information on batting events for the ECC is incomplete. This is not to diminish the very real accomplishments of ECC players. Indeed, 10 ECC position players were elected to the Baseball Hall of Fame. They played under extremely difficult conditions and may have been as good as, or even better than, some of the leading hitters in subsequent eras. The problem is that there is no way to tell for sure. We have to content ourselves with the knowledge that there were some exceptionally talented players during the ECC, but we will never know for sure how they would have performed if the rules had been closer to the rules of subsequent eras or if they had played in one of those subsequent eras.

❋ 2 ❋

The Live Ball Interval, 1893–1900

THE AGE OF ED DELAHANTY

Big Ed Delahanty was one of the greatest hitters in baseball history, but his accomplishments have been overshadowed by the circumstances of his unusual death. Today few fans are aware that he had a .346 career batting average, that he once hit four home runs in a single game, or that he is the only man ever to be credited with batting titles in both the National and American leagues. If they have heard of Delahanty at all, it is because he died by being swept over Niagara Falls.

The Baseball Encyclopedia, p. 278

The frequent changes in leagues, teams, and rules finally tapered off. In 1891 the American Association went bankrupt, leaving the National League as the only major league. The number of teams in the National League remained at 12 until 1900, and there were no changes among the teams themselves.[1] Perhaps most important of all was a hiatus in the number and significance of changes in the rules governing the relationship between pitchers and batters.[2]

The result was an eight-year period (1893–1900) in which batters prospered. Because it was too short to be called an era, but too long to be ignored, it can be called the Live Ball Interval. It constitutes a short batters' interval sandwiched between two longer pitchers' eras. The ball definitely came "alive," as is illustrated by the following National League statistics:

	ERPG	AVG	SLG	OPS	HR%
Era of Constant Change (1876–1892)	6.14	.253	.339	640	.6
Live Ball Interval (1893–1900)	8.33	.287	.383	736	.7
Dead Ball Era (1901–1920)	5.72	.255	.334	650	.5

The impact on pitchers and batters was immediate: in 1893 all five of these statistics increased; in 1894 they peaked and then gradually declined, resulting in the averages noted above for the interval as a whole. It is important to note that the Live Ball Interval did not produce a great increase in home runs like subsequent live ball eras. Slugging averages increased, but mostly as a result of increases in singles and triples. The ball was "alive" and jumping off the bat, but it wasn't leaving the ballparks.

The top 10 hitters of the Live Ball Interval for each of the nine average measures of hitting performance are identified in table 2.1. If you look at the top 10 as a group, there is general agreement on nine of the players that should be included. All nine measures agree on four players (Delahanty, Kelley, Tiernan, and Hamilton), and all except one measure agree on five other players (Joyce, Duffy, McGraw, Smith, and Burkett).

If you look at the specific rankings of the players within the top 10, however, there is considerable disagreement. Ed Delahanty appears to be the highest-ranked hitter, but he doesn't enjoy the unanimity that Dan Brouthers did in the Era of Constant Change. He is rated first in four models, third or fourth in four other models, and sixth in one model. Billy Hamilton is also ranked first in four models, but he is ranked second or third in three models and seventh and tenth in two models. Bill Joyce is ranked second by four models, but two other players are ranked second by five other models. Joyce is ranked fourth by one model and lower than 10th by another. Three other players—Duffy, Tiernan, and Burkett—are ranked third by one measure, but Tiernan is ranked as low as ninth by two measures and Duffy and Burkett are ranked lower than 10th by one measure each. The more you proceed, the more disagreement you encounter.

There is a special problem contributing to the amount of disagreement between the various measures in this interval. Most, if not

TABLE 2.1 The Top 10 Hitters in the LBI by Various Measures

Player	PRG	Advanced Weighted Measures			Basic Weighted Measures			Unweighted Measures	
		LSLR	RC/27	LWTS	TA	OPS	SLG	OBP	AVG
Ed Delahanty	1	3	4	3	4	1	1	6	1
Bill Joyce	2	2	3	4	3	2	2	3	—
Hugh Duffy	3	7	5	7	8	8	4	—	5
Joe Kelley	4	8	9	8	6	7	4	7	6
Mike Tiernan	5	6	5	5	5	6	3	9	9
Hugh Jennings	6	—	—	—	—	—	—	10	8
Billy Hamilton	7	1	1	1	1	3	10	2	2
Jake Beckley	8	—	—	—	—	—	8	—	—
John McGraw	9	4	2	2	2	4	—	1	4
Elmer Smith	10	—	8	9	9	9	9	8	10
Jimmy Ryan	(11)	10	—	—	—	10	7	—	—
Jesse Burkett	(13)	5	7	6	7	5	6	5	3
Cupid Childs	(14)	9	10	9	10	—	—	4	—
George Van Haltren	(16)	—	—	—	—	—	—	—	7

all, of the leading hitters in most historical eras are middle-of-the-order batters. In this interval, however, three of the leading hitters—Hamilton, McGraw, and Burkett—were leadoff batters. They will be rated as leadoff batters in chapter 10.

If one ranks the remaining players in table 2.1 as middle-of-the-order batters and compares their PRG and LSLR rankings, as was done in the previous era, the disagreements can be narrowed. The revised alternative pair of rankings are (PRG first/LSLR second): #1 Delahanty/Joyce, #2 Joyce/Delahanty, #3 Duffy/Tiernan, #4 Kelley/Duffy, #5 Tiernan/Kelley, #6 Jennings/Childs, #7 Beckley/Van Haltren, #8 Smith/Smith, #9 Davis/Hoy, and #10 Childs/Jennings. Who should be ranked number 1? Delahanty deserves the top ranking because his record was maintained over a much longer period of time (16 years and 8,000+ plate appearances) than that of Joyce (eight years and 4,000+ plate appearances).

A second question arises: should Duffy, Kelley, and Tiernan be ranked third, fourth, and fifth, respectively, based on the PRG measure, or should Tiernan, Duffy, and Kelley be ranked third, fourth, and fifth, respectively, based on the LSLR weighted measure? Duffy and Kelley are ranked higher by the PRG measure because they had more runs batted in from the bases than Tiernan. Jennings, Beckley, and Davis are ranked higher in PRG than Childs, Van Haltren, and Hoy because the former had more runs batted in from the bases than the latter. The difference between the two rankings is so small, however, that one might argue for either. The most accurate ranking of the players for this interval may lie somewhere between the two rankings.

Table 2.2 summarizes the leading hitters of the Live Ball Interval by the positions they played in the field. Three of the 10 leading hitters in the Era of Constant Change had been first basemen. In the Live Ball Interval, however, things were different. Three of the 10 leading hitters were left fielders—Ed Delahanty, Joe Kelley, and Elmer Smith. The highest-ranking first baseman, Jake Beckley, ranked only eighth.

The highest-ranking left fielder and the highest-ranking hitter overall was Ed Delahanty, with a PRG rating of 1.263. Delahanty played from 1888 to 1903. He was one of the least-traveled players of his time, having spent 13 of his 16 major league seasons with the Philadelphia Phillies and three other seasons split between two other teams. Delahanty was one of the taller players of his time (6 feet 1 inch) and was known as "Big Ed." Early in his career he struggled both in the infield and at the plate, but after he was transferred to the outfield both his fielding and hitting improved markedly.

Delahanty was both a percentage hitter and a slugger. He led his league in slugging average five times and had the highest slugging average for the entire interval. Delahanty also led the league in doubles five times and in runs batted in three times. He is the only player in the history of Major League Baseball to win a batting title in both the National League (1899) and the American League (1902). They were the only batting titles he ever won, but he played at a time when batting averages were very high, giving him a lot of

TABLE 2.2 Leading Hitters of the LBI by Position[1]

Position	Leading Hitter	Second-Leading Hitter and Others in the Top 10 Overall
Left field	Ed Delahanty — 1.263 (1)*	Joe Kelley — 1.148 (4)*;
		Elmer Smith — 1.083 (10)
Third base	Bill Joyce — 1.218 (2)	John McGraw — 1.084 (9)
Center field	Hugh Duffy — 1.150 (3)*	Billy Hamilton — 1.089 (7)*
Right field	Mike Tiernan — 1.099 (5)	Patsy Donovan — .861 (31)
Shortstop	Hugh Jennings — 1.096 (6)*	George Davis — 1.046 (12)*
First base	Jake Beckley — 1.089 (8)*	Jack Doyle — 1.031 (15)
Second base	Cupid Childs — 1.040 (14)	Tom Daly — 1.008 (18)
Catcher	Duke Farrell — 1.010 (16)	Heinie Peitz — .966 (23)

* Indicates that a player was elected to the Baseball Hall of Fame.

[1] The number after each player's name is his potential runs per game (PRG) rating, and the number in parentheses is his PRG ranking for the LBI.

competition. He had the highest batting average and slugging average for the interval as a whole. Delahanty batted over .400 three times, a feat later equaled only by Ty Cobb and Rogers Hornsby. He ranks fourth all-time in career batting average and second all-time among right-handed batters. Delahanty was a fast base runner—he stole 455 bases, had 186 triples, and in one game had four consecutive inside-the-park home runs. He was also one of the best-fielding outfielders of his time. Sadly, his career was cut short at 35 when he fell from a railroad bridge over the Niagara River and was swept over Niagara Falls. Delahanty was elected to the Baseball Hall of Fame by the Old-Timers Committee in 1945.

The highest-ranking third baseman and the second-highest-ranking hitter overall was Bill Joyce, with a PRG rating of 1.218. Joyce had a relatively short (eight years) yet well-traveled career (five teams). He played from 1890 to 1898: two-plus years each with the Washington Senators and the New York Giants and three years with three other teams. Joyce was known as "Scrappy Bill" for his very aggressive style of baseball. He fought with his opponents, the umpires, and the office personnel of his own teams. Joyce sat out one entire season (1893) because after injuring himself his team wanted him to take a pay cut the following year and he refused.

Joyce was a power-hitting third baseman with a high on-base percentage. He seldom led his league in an offensive event—home runs once and walks twice—but he was a steady player with a batting average over .300 four times, an on-base percentage over .400 six times, and a slugging average over .500 four times. His .467 slugging average was the second highest and his .435 on-base percentage was the third highest for this interval. Joyce was the first major league player to hit four triples in one game. In 1898 he became player-manager of the Giants but quit before the season ended because of a dispute with his front office about how to discipline some of his players. He was having a successful season but never returned to Major League Baseball as a manager or as a player. Joyce was never elected to the Baseball Hall of Fame, probably because he was so contentious and had such a short career.

The highest-ranking center fielder and the third-highest-ranking hitter overall was Hugh Duffy, with a PRG rating of 1.150. Duffy played from 1888 to 1906: nine years with the Boston Braves and eight years with five other teams. At 5 feet 7 inches and 168 pounds, he was small but was both a great fielder and a great hitter. For six years he shared the Boston outfield with Tommy McCarthy, and their defensive exploits earned them the nickname the "Heavenly Twins." Duffy led the league in RBI and doubles once each and in hits, home runs, and batting average twice each. In 1894 he batted .440, the best single-season batting average ever. His lifetime batting average of .326 ranks fifth all-time for center fielders and eleventh all-time for right-handed batters. Duffy was elected to the Baseball Hall of Fame by the Old-Timers Committee in 1945.

The second-highest-ranking left fielder and the fourth-highest-ranking hitter overall was Joe Kelley, with a PRG rating of 1.148. Kelley played from 1891 to 1906: six-plus years with the Baltimore Orioles (NL), four-plus years with the Cincinnati Reds, and six years with three other teams. He was both a percentage hitter and a slugger. He batted over .300 11 years in a row and drove in over 100 runs in five of those years. Kelley never led his league in any major batting event, but he did have one of the best days ever in 1894 when he had nine hits in nine at bats, including four singles, four doubles, and a triple. He was one of the best-fielding outfielders of his time and was a good base runner. Kelley was elected to the Baseball Hall of Fame by the Veterans Committee in 1971.

The highest-ranking right fielder and the fifth-highest-ranking hitter overall was Mike Tiernan, with a PRG rating of 1.099. Tiernan played from 1887 to 1899. He was one of the least-traveled players of his day: he spent his entire 13-year major league career with the New York Giants. An unassuming player, Tiernan was known as "Silent Mike." Like many of the leading hitters of this interval, he was both a percentage hitter and a slugger. Tiernan's single-season accomplishments are modest—leading his league in walks, runs, and slugging once each and in home runs twice. He had a lifetime batting average of .311 and a lifetime slugging average of .463, which was the third best for this interval. Tiernan has never been elected to the Baseball Hall of Fame.

The highest-ranking shortstop and the sixth-highest-ranking hitter overall was Hugh Jennings, with a PRG rating of 1.096. Jennings's career lasted 28 years, from 1891 to 1918, but he actually played in only 18 of those years: five years with the Baltimore Orioles, two years split between Baltimore and another team, and five years with four other teams. During his last six years he played in a total of only 12 games. Jennings's nickname was "Ee-yah," which referred to his enthusiastic yell when one of his teammates did something good on the ball field. As previously mentioned, Jennings and McGraw played together for the Orioles from 1893 to 1898. The Orioles were managed by the notorious Ned Hanlon, who combined scientific baseball, intimidation, and dirty play in a win-at-any-cost approach to the game. Jennings was more of a percentage hitter than slugger and was ideally suited to Hanlon's style of play.

Jennings's specialty was getting on base by being hit by the pitcher. He led the league in HP five years in a row from 1894 to 1898. In 1896 he was HP 49 times, a single-season record that lasted 75 years. Jennings never led his league in any major batting event, but he was a steady hitter with a .312 lifetime batting average, sixth all-time for shortstops. He was also an outstanding defensive player. The combination of his aggressiveness, fielding ability, and steady hitting made him one of the best all-around players in baseball. Jennings was elected to the Baseball Hall of Fame by the Old-Timers Committee in 1945.

The second-highest-ranking center fielder and the seventh-highest-ranking hitter overall was Billy Hamilton, with a PRG rating of 1.089. Hamilton played from 1888 to 1901: six years each with the Philadelphia Phillies and Boston Braves and two years with the Kansas City Cowboys (AA). He was a small (5 feet 6 inches and 165 pounds) left-handed-hitting center fielder. The combination of his ability to get on base frequently (he ranks fourth all-time in on-base percentage) and his ability to run the bases with reckless abandon (he was known as "Sliding Billy") made him a great leadoff batter (see chap. 10). Hamilton led the league in walks, on-base percentage, and stolen bases five times each and in runs scored four times. He holds two major league records for runs scored: the single-season record (198 in 1894) and the all-time career record (1.06 per game). His 1894 on-base percentage of .523 ranks sixth all-time. Hamilton was elected to the Baseball Hall of Fame by the Veterans Committee in 1961.

The highest-ranking first baseman and the eighth-highest-ranking hitter overall was Jake Beckley, with a PRG rating of 1.089. Beckley played from 1888 to 1907: eight years with the Pittsburgh Pirates, seven years with the Cincinnati Reds, and five years with three other teams. The nickname "Jake" was short for Jacob. He was also called "Eagle Eye" because he was so good at telling the difference between good and bad pitches. Only once did Beckley lead his league in a major batting event—doubles in 1890—but he batted over .300 in 13 seasons for a lifetime average of .308. Beckley ranks fourth all-time in triples (244). He was not a particularly good fielder and had a weak throwing arm, but he was very durable—ranking second all-time for games played at first base and first all-time in putouts by a first baseman. Beckley was elected to the Baseball Hall of Fame by the Veterans Committee in 1971.

The second-highest-ranking third baseman and the ninth-highest-ranking hitter overall was John McGraw, with a PRG rating of 1.084. McGraw played from 1891 to 1906: eight seasons with the Baltimore Orioles (NL) and eight seasons with four other teams, including the Orioles when they played in the American Association and American League. McGraw and Hugh Jennings played together for the Baltimore Orioles from 1893 to 1898. McGraw succeeded Ned Hanlon as player-manager of the Orioles and a few years later began a long and illustrious career as manager of the New York Giants. Hanlon's aggressive approach to the game resonated with McGraw, who was nicknamed "Little Napoleon." McGraw, like Jennings, was good at getting on base by being hit by the pitcher, but his specialty was fouling off pitches. Long after retiring as a player, he purposely fouled off 26 pitches in a row in a spring training game. McGraw led the league in walks and runs twice, in on-base percentage three times, and had the highest on-base percentage for this interval as a whole. He never led the league in stolen bases, but he was one of the fastest base runners of his time. McGraw was a good, but not a great, fielding third baseman. His lifetime batting average of .334 and on-base percentage of .466 rank first all-time for third basemen. McGraw was elected to the Baseball Hall of Fame in 1937 by the Centennial Commission, but as a manager and not as a player.

The third-highest-ranking left fielder and the 10th-ranked hitter overall was Elmer Smith, with a PRG rating of 1.083. Smith played from 1886 to 1901: six years with the Pittsburgh Pirates, four years with the Cincinnati Red Stockings (AA), three years with the Cincinnati Reds, and a year split between two other teams. He started out as a successful pitcher but developed a sore arm and was sent to the minor leagues. After two years Smith returned to the majors, but as a successful outfielder. He was fast and his arm was rejuvenated. Smith used a big 54-ounce bat but hit only 37 career home runs. He was more of a percentage hitter than a slugger. He batted over .300 six years in a row and had a lifetime batting average of .310. Smith has never been elected to the Baseball Hall of Fame.

The highest-ranking second baseman, Cupid Childs, ranked 14th overall with a PRG rating of 1.040. Childs played from 1888 to 1901: eight years with the Cleveland Spiders (NL) and five years with four other teams. His first name was Clarence; why he was called "Cupid" is purely speculative. Childs was probably the best-fielding second baseman of his time. For several years he led the league in various defensive categories and holds the all-time major league record for the number of chances at second base in a nine-inning game (18). He led the league once each in runs, doubles, and on-base percentage and had the third best on-base percentage for this interval as a whole. Childs has never been elected to the Baseball Hall of Fame.

The highest-ranking catcher, Duke Farrell, ranks 16th overall with a PRG rating of 1.010. Farrell played from 1888 to 1905 and was one of the most-traveled players of his day. He spent four years with the Brooklyn Dodgers, three-plus years with the New York Giants, three-plus years with the Washington Senators (NL), and seven years with five other teams. Farrell's first name was Charles, but he was called "Duke" because he came from Marlboro (the name of a famous British Duke), Massachusetts. He was a switch-hitter and led the American Association in home runs and RBIs one year but never again led his league in any other major batting event. His greatest asset was his throwing arm. One day in 1897 Farrell threw out eight of nine Baltimore Orioles attempting to steal on him. This is an all-time major league record. Farrell has never been elected to the Baseball Hall of Fame.

Eight position players of the Live Ball Interval were named to the Baseball Hall of Fame. Six of them—Ed Delahanty, Hugh Duffy, Joe Kelley, Hugh Jennings, Jake Beckley, and Billy Hamilton—ranked in the top 10 in PRG. Two others—George Davis and Jesse Burkett—ranked 12th and 13th, respectively. Jennings and Davis were regarded as the best all-around shortstops of their time, and fielding had to have been a factor in their election. Four position players ranked in the top 14 in PRG—Bill Joyce (second), Mike Tiernan (fifth), Elmer Smith (10th), and Cupid Childs (14th)—were not elected to the Hall of Fame. Joyce had a short career and was not a particularly good fielding third baseman. Tiernan and Smith were good hitters but were not particularly fast and did not distinguish themselves in the field. John McGraw was an excellent leadoff batter but had a relatively short playing career. His managing career was much longer and so successful that he was elected to the Hall of Fame as a manager.

How do the hitters of the Live Ball Interval compare with the hitters of other eras? In chapter 1 it was shown that the pitcher-batter rules of baseball during the Era of Constant Change were so different from the rules of subsequent eras that a meaningful comparison of hitters in that era with hitters in subsequent eras was not possible. By 1891 pitcher-batter rules had evolved considerably toward what they are today, but one very significant difference remained. From 1892 to 1900, home plate was still a four-sided 12-inch square, giving batters an advantage over pitchers not enjoyed in subsequent eras. In 1901, home plate was changed to its present five-sided, 17-inch-wide shape. There was another important difference in the way baseball was played in the first two eras of the game. In the ECC, an average of 4.6 errors were committed in each game (both teams), and in the LBI the average was still high—3.6 errors per game. In the DBE, the rate declined to 2.1 per game, and since then it has averaged 1.1 per game.

Although the conditions in the LBI were closer to those of today than the conditions in the ECC, the much smaller home plate and the high rate of errors make it difficult to compare the hitters of that era with the hitters of subsequent eras. There were great hitters in the LBI—eight LBI position players were elected to the Hall of Fame based mainly on their hitting prowess. But, as is the case with the ECC, we don't know for sure how they compare with hitters of subsequent eras because the conditions were so different.

3

The Dead Ball Era, 1901–1920
THE AGE OF TY COBB

It wasn't that he was so fast on his feet, although he was fast enough. There were others who were faster . . . It was that Cobb was so fast in his thinking. He didn't outhit the opposition and he didn't outrun them. He outthought them.

Sam Crawford, teammate of Ty Cobb

The year 1901 marked the beginning of a period of remarkable stability for Major League Baseball. Except for the very brief appearance of the Federal League (1914 and 1915), the number of major leagues remained at two—the National League and the American League—for the 108 years from then to the present time. The number of teams in each league remained at eight for 60 years, until in 1961 the American League expanded to 10 teams. Strict enforcement of the reserve clause severely restricted the movement of players between teams.[1] The frequent changes in the rules governing the relationship between pitchers and batters had already tapered off, and by 1901 the rules were not much different from what they are today. The infrequent changes since that time have pertained to spitball pitching, the height of the pitcher's mound, the dimensions of the strike zone, and the designated hitter rule.

The Dead Ball Era was a pitcher's era for several reasons. First, the 1900 expansion of home plate from a 12-inch square to a five-sided figure 17 inches wide favored pitchers over batters. Second, the spitball was permitted as a legal pitch. Third, the ball itself was literally dead. Game balls were kept in play until they were lost. If balls were hit into the stands, the ushers would retrieve them. If balls were hit out of sight, they would search and return them to play. As a result, the game ball quickly became old and dirty. It wasn't until 1920 that the spitball rule was changed and worn baseballs were regularly replaced by new ones, thus facilitating an end to the Dead Ball Era.

The period from 1901 to 1920 was much different from the period before it and the period after it. Consider the following National League statistics:

	ERPG	AVG	SLG	OPS	HR%
Live Ball Interval (1893–1900)	8.33	.287	.383	736	.7
Dead Ball Era (1901–1920)	5.72	.255	.334	650	.5
Live Ball Era (1921–1941)	7.90	.280	.393	730	1.4

Live Ball Interval statistics are not available for the American League because it wasn't organized until 1901, but the differences in the American League statistics between the Dead Ball Era and the Live Ball Era are similar to the differences between the two eras in the National League. The Dead Ball Era was quite clearly a period in which pitchers prevailed over batters.

The top 10 hitters of the Dead Ball Era for each of the average measures of hitting performance are identified in table 3.1. The pattern of agreement and disagreement between the measures is similar to the pattern in the two preceding eras. If you look at the top 10 as a group, there is general agreement that eight players belong there. All of the measures agree on five players (Cobb, Jackson, Speaker, Wagner, and Donlin), all except one measure agree on Lajoie, and all except two measures agree on two other players (Collins and Clarke). Sixteen players share the top 10 rankings (about the same as for the two preceding eras), and 10 players share the top five rankings (a little more than in the two preceding eras).

If you look at the specific rankings of the players within the top 10, there is more disagreement. All except one measure agree that Ty Cobb is first. Joe Jackson is ranked second by five measures, but two other players are ranked second by four other measures. Tris Speaker is ranked third by six measures. Cobb, Jackson, and Speaker constitute the top three in all measures except one, and that one includes

TABLE 3.1 The Top 10 Hitters in the DBE by Various Measures

Player	PRG	Advanced Weighted Measures			Basic Weighted Measures			Unweighted Measures	
		LSLR	RC/27	LWTS	TA	OPS	SLG	OBP	AVG
Ty Cobb	1	1	1	1	1	1	2	1	1
Joe Jackson	2	2	2	3	3	2	1	4	2
Tris Speaker	3	3	3	2	2	3	3	2	3
Gavy Cravath	4	—	—	—	8	4	4	—	—
Honus Wagner	5	4	4	4	5	4	6	6	8
Nap Lajoie	6	9	8	8	10	8	7	—	5
Bobby Veach	7	—	—	—	—	—	—	—	—
Sam Crawford	8	10	—	—	—	10	8	—	—
Home Run Baker	9	—	—	—	—	—	10	—	—
Mike Donlin	10	8	5	6	7	6	5	8	6
Eddie Collins	(12)	7	6	5	4	7	—	3	6
Zach Wheat	(13)	—	—	—	—	9	9	—	9
Frank Chance	(15)	—	7	9	6	—	—	5	—
Fred Clarke	(16)	6	9	7	9	10	—	8	10
Roger Bresnahan	(21)	—	—	—	—	—	—	8	—
Willie Keeler	(32)	5	10	10	—	—	—	7	4

Cobb and Speaker in the top three but ranks Jackson fourth. Honus Wagner is ranked fourth, fifth, or sixth by all measures except one. As you proceed further, there is more and more disagreement.

For the first two historical eras the differences in the rankings of the various measures were approached by comparing the PRG rankings with the LSLR rankings, the most reliable measure actually paired with runs scored and tested for that period of time. The team PRG data for those eras had not been paired with runs scored and tested because the conditions under which the game was played, especially in the Era of Constant Change, were radically different from those in subsequent eras. There was also a very high rate of errors, and the data were incomplete. Those conditions had dissipated by the time of the Dead Ball Era. The PRG team data for the Dead Ball Era were included in the statistical tests that indicated that the

PRG model was the best measure of team performance from 1901 to 2009.

There is no proof that the player PRG measure is better than the other measures, but it can be argued that it is better because the player PRG measure was extrapolated from the team PRG measure, which had tested best. And the margin of primacy in those team tests should be more than sufficient to offset any loss of accuracy in the player extrapolations. The PRG and LSLR rankings for this era, by the way, are not that far apart. They agree on the specific ranking of the first three players, with Cobb first, Jackson second, and Speaker third. They also agree that Wagner should be fourth or fifth and Lajoie sixth or ninth. The biggest disagreement concerns Cravath, Collins, and Clarke. Cravath ranks higher in PRG than in LSLR, and Collins and Clarke rank lower in PRG than in LSLR. Cravath ranks first in the era for RBI per at bat, and Collins and Clarke rank very low in RBI per at bat. Runs batted in are an important element in the PRG system of measurement.

The differences between the two measures are not great, are explicable, and do not have to be accounted for because this comparison is only an aside, albeit an interesting aside. It cannot be proven that any measure, PRG included, is the best measure of player performance, but the evidence suggests, as noted above, however, that the PRG measure may be the best.

Table 3.2 summarizes the leading hitters of the Dead Ball Era by the positions they played in the field. This was an era in which hitting was dominated by outfielders—7 of the 10 leading hitters were outfielders. Center fielders led the outfielders—3 of the 10 leading hitters were center fielders. The highest-ranking center fielder and the highest-ranking hitter overall was Ty Cobb, with a PRG rating of 1.215. Cobb played from 1905 to 1928 and spent all but the last two years of his 24-year major league career with the Detroit Tigers. He was nicknamed the "Georgia Peach" because he came from Georgia and was a great player. Cobb was not a peach of a person—he was temperamental, violent, and often in trouble—but he was a peach of a player.

Ty Cobb's greatest accomplishment was his lifetime batting aver-

TABLE 3.2 Leading Hitters of the DBE by Position[1]

Position	Leading Hitter	Second-Leading Hitter and Others in the Top 10 Overall
Center field	Ty Cobb — 1.215 (1)*	Tris Speaker — 1.163 (3)*; Mike Donlin — 1.073 (10)
Left field	Joe Jackson — 1.190 (2)	Bobby Veach — 1.099 (7)
Right field	Gavvy Cravath — 1.145 (4)	Sam Crawford — 1.085 (8)*
Shortstop	Honus Wagner — 1.140 (5)*	Bill Dahlen — .984 (20)
Second base	Nap Lajoie — 1.128 (6)*	Eddie Collins — 1.044 (12)*
Third base	Home Run Baker — 1.075 (9)*	Heinie Zimmerman — .999 (17)
First base	Charlie Hickman — 1.037 (13)	Frank Chance — 1.029 (15)*
Catcher	Roger Bresnahan — .974 (21)*	Hank Severeid — .918 (33)

* Indicates that a player was elected to the Baseball Hall of Fame.
[1] The number after each player's name is his potential runs per game (PRG) rating, and the number in parentheses is his PRG ranking for the DBE.

age of .367, the all-time major league record. He also ranks high in several other all-time hitting events: second in hits, runs, and triples; fourth in doubles, total bases, and stolen bases; fifth in runs batted in; eighth in on-base percentage; and 10th in extra-base hits. Curiously, he does not hold a single-season record in any event. He did, however, lead the American League in batting average 12 times, hits and slugging average eight times, on-base percentage six times, runs five times, triples and runs batted in four times each, and doubles three times.

Ty Cobb led the league in home runs only once, with nine in 1909, and hit only 117 home runs in his entire career. This was not just because of the influence of the "dead ball" but also a conscious choice on his part. Ty Cobb tried to get base hits rather than home runs. One day he proved to a reporter that he could hit home runs if he wanted to. Instead of choking up on the bat with his hands apart, he held his hands together at the end of the bat. Instead of trying to get hits, he tried to hit home runs. The result was three home runs. To prove that it was no coincidence, he hit two more home runs the following day!

Cobb was also a great base runner and fielder. He led his league in stolen bases six times and ranks fourth all-time in stolen bases. Cobb ranks second all-time for assists and double plays by an outfielder and fifth all-time in putouts by an outfielder. Ty Cobb was the very first player to be elected to the Baseball Hall of Fame (1936). Players from previous eras who were elected to the Hall of Fame were elected after 1936 by either the Old-Timers Committee or the Veterans Committee.

The highest-ranking left fielder and the second-highest-ranking hitter overall was Joe Jackson, with a PRG rating of 1.190. Jackson played from 1908 to 1920: four-plus years with the Cleveland Indians, four-plus years with the Chicago White Sox, and four years in which he played in a total of only 47 games. Early in Jackson's career he played a game in his stocking feet because a new pair of baseball shoes had given him blisters. From that time on he was known as "Shoeless Joe."

Jackson was both a percentage hitter and a slugger. He ranked first in slugging average and second in batting average for his era. He hit only 54 home runs, but he had a lot of doubles and triples. Jackson led the American League in triples three times, hits twice, and doubles, on-base percentage, and slugging average once each. His lifetime batting average was .356, the third highest all-time. Jackson was only 31 years old in his last season; what his batting average would have been had he played longer is problematic. It is doubtful, however, that he could have overtaken Cobb (.367), the all-time leader. Jackson hit .382 in his last season. But even if he had played five more seasons and hit .382 for every one of those seasons, his lifetime batting average would still have been lower than Cobb's.

A lifetime batting average of .356, however, is not bad. Jackson doesn't hold any single-season records, but he does rank 15th all-time in on-base percentage. The story of Shoeless Joe Jackson is one of the great tragedies in baseball history. He confessed to conspiracy to throw the 1919 World Series and, along with seven teammates, was brought to trial but acquitted by the jury. Nevertheless, the commissioner of baseball, Kenesaw Mountain Landis, banned all eight players from baseball for life. Thus, Jackson has never been elected to the Baseball Hall of Fame.

The second-highest-ranking center fielder and the third-highest-ranking hitter overall was Tris Speaker, with a PRG rating of 1.163. Speaker played from 1907 to 1928: 11 years with the Cleveland Indians, nine years with the Boston Red Sox, and two years with two other teams. Tris was a shortened version of Tristam, his first name. His nickname was the "Grey Eagle" for the way in which he patrolled center field.

Speaker led his league in doubles eight times, on-base percentage four times, hits twice, and slugging average and home runs once each. Speaker's lifetime batting average of .345 is the fifth highest all-time and third highest for left-handed batters. He holds the all-time career record for doubles (792) and is fifth in hits (3,514), sixth in triples (222), 10th in runs, and 11th in on-base percentage and extra-base hits. Only once did Speaker lead the league in batting average because he was a contemporary of Ty Cobb, who led the league 12 times.

Speaker's fielding accomplishments are also impressive. He holds the all-time record for career assists and double plays for an outfielder. Speaker was able to make so many assists and double plays because he played a very shallow center field. It is not surprising that Speaker could get away with that in the 1,769 games he played in the Dead Ball Era, but he also got away with it, although at a lower rate, for the 1,020 games he later played in the Live Ball Era. His secret, apparently, was an uncanny ability to get a jump on the ball as it was hit. Speaker also ranks second all-time for putouts by an outfielder. He was certainly one of the greatest all-around center fielders ever. Speaker was elected to the Baseball Hall of Fame in 1937.

The highest-ranking right fielder and the fourth-highest-ranking hitter overall was Gavvy Cravath, with a PRG rating of 1.145. Cravath played from 1908 to 1920: nine years with the Philadelphia Phillies and two years split among three other teams. Gavvy was a nickname given to Cravath by the baseball world. It apparently came from the Spanish word *gaviotas*, meaning "seagull," but it is not apparent why it was applied to Cravath. He was also called "Cactus" for the prickly side of his personality.

Cravath was more of a slugger than a percentage hitter—the opposite of most Dead Ball Era hitters. He ranked fourth in slugging

average, but below 10th in batting average. A right-handed oppo-
site-field power hitter, he was helped by the short right-field fence in
Philadelphia's Baker Bowl. Cravath holds no career or single-season
hitting records, but he did lead the league in home runs six times, on-
base percentage and slugging average twice each, and hits and runs
once each. He had a career batting average of only .287, but his 119
home runs was more than any other player in the Dead Ball Era. Cravath has never been elected to the Baseball Hall of Fame.

The highest-ranking shortstop and the fifth-highest-ranking
hitter overall was Honus Wagner, with a PRG rating of 1.140. He
played from 1897 to 1917: 18 years with the Pittsburgh Pirates and
three years with the Louisville Colonels (NL). Wagner's first name
was John. His nickname, Honus, came from the German word
for John—Hans or Johannes. Wagner was also known as "the Fly-
ing Dutchman." Personality-wise, he was quiet, unassuming, and
friendly—the opposite of Ty Cobb. Baseball-wise, however, Wagner
was the same as Cobb—a dominant force in his league.

Honus Wagner was an all-around shortstop. He was an excellent
fielder—quick with a strong arm. Wagner was also an excellent base
runner—ranking 10th all-time in stolen bases. His career batting average of .328 ranks only 32nd overall, but it ranks first among all
shortstops, a position filled with fielding ability in mind more than
hitting ability.

Wagner led the National League in batting average eight times,
doubles seven times, slugging average six times, runs batted in five
times, on-base percentage four times, triples three times, and hits
and runs twice. He hit very well against legends Cy Young, Christy
Mathewson, and other great pitchers of that era. Wagner ranks high
in several all-time hitting events: third in triples, sixth in hits, eighth
in doubles, 18th in runs batted in, 19th in runs, and 20th in total
bases—unusually high rankings for a shortstop, especially one who
played in the pitcher-dominated Dead Ball Era. Wagner was elected
to the Baseball Hall of Fame in 1936. Ty Cobb, Babe Ruth, and Honus Wagner were the first three position players chosen.

The highest-ranking second baseman and the sixth-highest-ranking hitter overall was Nap Lajoie, with a PRG rating of 1.128. Lajoie

played from 1896 to 1916: 13 years with the Cleveland Indians, five years with the Philadelphia Phillies, and three years with the Philadelphia Athletics. Lajoie and Eddie Collins were the dominant second basemen of the era. Collins may have been a little better in the field—he had the highest fielding average nine times to Lajoie's six times—but Lajoie was a better hitter. Lajoie was both a slugger and a percentage hitter, ranking seventh in slugging average and fifth in batting average, but Collins was only a percentage hitter, ranking sixth in batting average.

Nap is short for Napoleon, his given name. He was also called Larry, short for his family name, Lajoie. He led the league in doubles five times; hits four times; runs batted in, batting average, and slugging average three times each; on-base percentage twice; and runs and home runs once each. Lajoie ranks sixth all-time in doubles, 12th in hits, and 19th in batting average. Nap Lajoie was elected to the Baseball Hall of Fame in 1937.

The second-highest-ranking left fielder and the seventh-highest-ranking hitter overall was Bobby Veach, with a PRG rating of 1.099. Veach played from 1912 to 1925: 12 years with the Detroit Tigers, followed by one year with the Boston Red Sox and one year split between Boston, the New York Yankees, and the Washington Senators. Veach was a left-handed-hitting outfielder with speed and a good arm. Like most of the leading hitters of his time, he hit for average (.310 lifetime) and not for power (only 64 career home runs). Veach hit over .300 in 10 of his 14 seasons. He led the American League in runs batted in three times, doubles twice, and hits and triples once each. His best year was 1919, in which he led the league in hits, doubles, and triples and posted a .355 batting average. In 1925 he had the unique experience of pinch-hitting for Babe Ruth. Yankee manager Miller Huggins was upset with Ruth and wanted to teach him a lesson. Veach was never elected to the Baseball Hall of Fame, probably because he had to compete with so many big-name players—Cobb, Jackson, Speaker, Keeler, Lajoie, Eddie Collins, Donlin, Wagner, Wheat, and Flick—with higher lifetime batting averages.

The second-highest-ranking right fielder and the eighth-highest-ranking hitter overall was Sam Crawford, with a PRG rating of 1.085.

Crawford played from 1899 to 1917: 15 years with the Detroit Tigers and four years with the Cincinnati Reds. During his years with Detroit, he played in some of the greatest outfields in the history of Major League Baseball—three years with Marty McIntyre and Ty Cobb, four years with Davey Jones and Ty Cobb, and four years with Bobby Veach and Ty Cobb. Crawford was a great fielder—he was fast and had a powerful arm. He ran the bases well and was a good hitter. He was known as "Wahoo Sam" because he grew up in Wahoo, Nebraska. An appropriate alternative nickname might have been "Mr. Triple" because he holds the all-time major league record for triples (309). Crawford led the American League in triples six times and had 10 or more triples in 17 straight seasons. He also led his league in runs batted in three times and in runs, doubles, and home runs once each. Crawford was elected to the Baseball Hall of Fame in 1957 by the Veterans Committee largely as the result of the efforts of his former teammate, Ty Cobb.

The highest-ranking third baseman and the ninth-highest-ranking hitter overall was Frank "Home Run" Baker, with a PRG rating of 1.075. Baker played from 1908 to 1920: seven seasons with the Philadelphia Athletics and six with the New York Yankees. He was given the nickname "Home Run" because he hit crucial home runs against Rube Marquard and Christy Mathewson on consecutive days in the 1911 World Series.

"Home Run" Baker was also a good-fielding third baseman. He ranks ninth all-time for chances accepted per game and seventh all-time for putouts per game for a third baseman. But, as his nickname implies, Baker was known primarily as a slugger. He was always a deep threat, because he was a pull hitter and carried a 52-ounce bat. The short right-field fence in Philadelphia's Baker Bowl made him even more of a threat at home. During his years with the Athletics, Baker was the leading home run hitter in the American League, leading the league in home runs four years in a row, from 1911 to 1914. He also led the league in runs batted in twice and triples once. Baker was elected to the Baseball Hall of Fame by the Veterans Committee in 1955.

The third-highest-ranking center fielder and the 10th-ranked hitter overall was Mike Donlin, with a PRG rating of 1.073. Donlin played from 1899 to 1914: a turbulent career in which he missed four entire seasons and played in 100 or more games only five times. His frequent absences were for different reasons—because of a jail sentence, an injury, a salary dispute, and the pursuit of a second career in acting. Donlin was one of the best hitters of his time, with a lifetime batting average of .333 (sixth for his era), but he may have been even better had it not been for the significant amount of downtime. He was pugnacious both on and off the field and, because he strutted like a turkey, was known as "Turkey Mike." Donlin has never been elected to the Baseball Hall of Fame.

The highest-ranking first baseman and the 13th-ranked hitter overall was Charlie Hickman, with a PRG rating of 1.037. Hickman played from 1897 to 1908. He was well traveled, playing for eight different teams during his 12-year major league career. Hickman played the longest for Washington (415 games) and Cleveland (319 games). He led the league in hits once but never in any other major batting event. He was nicknamed "Piano Legs" and was especially slow on the base paths. Hickman was a good but not a great player and has never been elected to the Baseball Hall of Fame.

The highest-ranking catcher, but only the 21st-ranked hitter overall, was Roger Bresnahan, with a PRG rating of .974. Bresnahan played from 1897 to 1915: six-plus years with the New York Giants, four years with the St. Louis Cardinals, three years with the Chicago Cubs, and three-plus years with two other teams. He was called the "Duke of Tralee" because his family had migrated to the United States from Tralee, Ireland.

Bresnahan played every position in the field, including pitcher, but he was a catcher 60% of the time. He was a fiery character who had many confrontations with umpires. Bresnahan was a great catcher and was fast enough to be used occasionally as a leadoff hitter, a rarity for a catcher. His batting credentials, however, are not impressive. He holds no all-time or single-season records and led the league in a major batting event only twice—in walks and on-base percentage. Bresnahan introduced three new pieces of equipment for catchers: shin guards, the padded face mask, and the batting helmet. The first two caught on quickly, but the batting helmet was an idea whose time didn't arrive until many years later. He was regarded by many as the preeminent catcher of his time. Bresnahan was elected to the Baseball Hall of Fame by the Old-Timers Committee in 1945.

Position players are elected to the Hall of Fame based on all of their accomplishments, but hitting is probably the biggest factor, and table 3.2 illustrates this point. Nine of the 19 position players of this era who were elected to the Hall of Fame are listed in table 3.2. Seven of the nine are ranked in the top 12 for the era and were undoubtedly elected based primarily on their hitting. Roger Bresnahan was a respectable hitter, but his election to the Hall of Fame was probably based more on his work behind the plate than his hitting accomplishments. Frank Chance may have been the only first baseman ever elected based primarily on his fielding. Three of the nine not listed in table 3.2—Willie Keeler, Elmer Flick, and Zach Wheat—were probably elected based more on their hitting than their fielding. Keeler is famous for his ability to "Hit Em Where They Ain't" and had a lifetime batting average of .341. Wheat and Flick were very good Dead Ball Era hitters, with lifetime batting averages of .317 and .313, respectively, and slugging averages of .450 and .445, respectively, both higher than Home Run Baker's slugging average of .442. Five of the nine not listed in table 3.2—Jimmy Collins, Johnny Evers, Harry Hooper, Ray Schalk, and Bobby Wallace—were great fielders and were undoubtedly elected based on their fielding accomplishments. Fred Clarke was probably elected based on a combination of his hitting and fielding. Joe Tinker, like his partners in the Tinker to Evers to Chance trio, was also elected based primarily on his fielding.

* 4 *

The Live Ball Era, 1921–1941
THE AGE OF BABE RUTH

Babe Ruth was . . . the greatest baseball player who ever lived. . . . His gargantuan appetites and prodigious talents . . . made him one of the most recognizable figures in American history. . . . His career also typifies the classic "rags to riches" scenario, in that he rose from the harsh milieu of a Baltimore reform school to achieve wealth, celebrity, and lasting renown. Ruth revolutionized the game with his unprecedented slugging, and in the wake of the "Black Sox Scandal" of 1919 he single handedly restored America's love of baseball.

Total Baseball, Chapter 5: "The Top 100 Players," by David Pietrusza, Matthew Silverman, Michael Gershman, and Mikhail Horowitz (2001)

A series of events in 1919 and 1920 set the stage for a radical transformation of Major League Baseball. During the 1919 season, Babe Ruth established a new record for home runs (29); other players followed his lead, and home runs and runs per game increased dramatically. The fans liked it and attendance (per game) also increased significantly. In the 1919 World Series underdog Cincinnati upset highly favored Chicago, and rumors spread that Chicago players had deliberately tried to lose. Then in December the Boston Red Sox sold Babe Ruth to the New York Yankees.

Before the 1920 season Major League Baseball became concerned with the potential danger to batters from doctored baseballs. They were also concerned with the threat to offensive baseball and attendance. The spitball and the scuffing of baseballs were prohibited. Each team was allowed, however, to name two spitball pitchers for the 1920 season only.[1] During that season, home runs, runs per game, and attendance were all higher than the year before. On August 16 Ray Chapman was hit by a pitch from Carl Mays and died the following day, the first on-the-field fatality in the history of Major League Baseball. The public outcry put pressure on umpires to discard all old and defaced baseballs and keep only clean balls in the game.

In September 1920, eight Chicago White Sox players were indicted for attempting to fix the 1919 World Series and details of the affair became public knowledge. Team owners were worried about the public reaction and sought the restoration of confidence in the integrity of the game. In November, they hired judge Kenesaw Mountain Landis as the first commissioner of Major League Baseball and gave him absolute power to act on their behalf.

During the 1921 season, home runs and runs per game were higher, but American League attendance declined significantly. In August, the accused Chicago White Sox players were acquitted of fixing the 1919 World Series. The very next day, however, a dubious Commissioner Landis banned all eight of them from baseball for life. Public confidence was restored, but it took time. Home runs and runs per game continued to increase, but it wasn't until 1924 that American League attendance rebounded to the 1920 level.

The result of this series of events was the end of one era and the beginning of another. The distinction between the Dead Ball Era and the Live Ball Era is vividly portrayed by the following statistics:

	AVG	OBP	SLG	OPS	HR%	ERPG
Dead Ball Era (1901–1920)	.255	.317	.334	651	.5	5.75
Live Ball Era (1921–1941)	.281	.344	.398	742	1.4	8.26
Live Ball Continued (1942–1962)	.259	.331	.382	713	2.2	7.66

There is also a clear and consistent distinction between the two live ball eras. All statistics for the 1942–1962 period except HR% are lower than the statistics for the 1921–1941 period, and all are higher than the statistics for the Dead Ball Era. The decline in offensive statistics in the 1942–1962 period was precipitated by the departure of many leading hitters during World War II. Offense rebounded when those hitters returned, but, as the statistics show, the level of offense from 1942 to 1962 was significantly lower than from 1921 to 1941.

The top 10 hitters of the Live Ball Era for each of the average measures of hitting performance are identified in table 4.1. The pattern of agreement and disagreement between the measures is similar to the pattern in the preceding eras. If you look at the top 10 as a group, there is general agreement on eight players who belong in the group. All of the measures agree on three players (Ruth, Gehrig, and Hornsby), all except one measure agree on two players (Greenberg and Foxx), and all except two measures agree on three other players (Heilmann, Ott, and Averill). Twenty-one players share the top 10 rankings (compared with from 14 to 16 in the previous eras), and nine players share the top five rankings (compared with from eight to 10 in the previous eras).

If you look at the specific ranking of the players within the top 10, there is some disagreement, but less so than in any of the previous eras. All except one measure agree that Babe Ruth is number 1 and Lou Gehrig is number 2; six measures agree that Rogers Hornsby is number 5; and five measures agree that Jimmie Foxx is number 3, Mel Ott is number 6, and Harry Heilmann is number 8. There is more disagreement, some of it substantial, with regard to the seven remaining players.

There may be less disagreement on the specific ranking of the players of this era than the previous eras, but that disagreement does, nevertheless, have to be dealt with. If you want to know who the best hitters of this era were, you have to pick one of these measures, or combine some of these measures, or have a measure of your own on which to base your ranking. How do you decide which it will be? Your answer to this question should be based on the guidance provided in the Pregame Analysis.

Table 4.2 summarizes the leading hitters of the Live Ball Era by the positions they played in the field. The highest-ranking right fielder and the highest-ranking player overall was Babe Ruth, with a PRG

TABLE 4.1 The Top 10 Hitters in the LBE by Various Measures

Player	PRG	Advanced Weighted Measures			Basic Weighted Measures			Unweighted Measures	
		LSLR	RC/27	LWTS	TA	OPS	SLG	OBP	AVG
Babe Ruth	1	1	1	1	1	1	1	1	2
Lou Gehrig	2	2	2	2	2	2	2	2	6
Hank Greenberg	3	3	4	3	5	4	4	7	—
Jimmie Foxx	4	4	3	3	3	3	3	4	—
Rogers Hornsby	5	5	5	5	4	5	5	3	1
Hack Wilson	6	—	10	7	7	7	7	—	—
Al Simmons	7	7	—	—	—	—	9	—	8
Harry Heilmann	8	—	8	9	8	8	—	8	2
Mel Ott	9	8	6	6	6	6	—	6	—
Earl Averill	10	6	7	8	9	9	10	—	—
Chuck Klein	(12)	—	—	10	—	10	8	—	—
Mickey Cochrane	(27)	—	9	—	10	—	—	5	—
Riggs Stephenson	(27)	—	—	—	—	—	—	9	7
Other players[1]	—	—	—	—	—	—	—	—	—

[1] Eight other players ranked in the top 10 but in one measure only: Bill Terry #4 in AVG, George Sisler #5 in AVG, Hal Trosky #6 in SLG, Bob Johnson #9 in LSLR, Paul Waner #9 in AVG, Charlie Gehringer #10 in LSLR, Heinie Manush #10 in AVG, and Arky Vaughn #10 in OBP.

rating of 1.480. Ruth played from 1914 to 1935: six years with the Boston Red Sox, 15 years with the New York Yankees, and one year with the Boston Braves. Ruth played minor league ball with the Baltimore Orioles. On the very first day of minor league spring training, the 19-year-old George Herman Ruth was referred to as the owner's "new babe." The nickname stuck, and he was thereafter referred to as the Babe. Many people, even serious baseball fans, can't tell you his real name. Babe Ruth was also referred to as the "Bambino," Italian for "baby," and the "Sultan of Swat," for his great slugging ability.

The Orioles sold Ruth to the Boston Red Sox, and he quickly developed into one of the best left-handed pitchers in the game. While with Boston he won 94 games and lost 46 for a winning percentage

TABLE 4.2 Leading Hitters of the LBE by Position[1]

Position	Leading Hitter	Second-Leading Hitter and Others in the Top 10 Overall
Right field	Babe Ruth — 1.480 (1)*	Harry Heilmann — 1.233 (8)*; Mel Ott — 1.223 (9)*
First base	Lou Gehrig — 1.429 (2)*	Hank Greenberg — 1.381 (3)*; Jimmie Foxx — 1.366 (4)*
Second base	Rogers Hornsby — 1.283 (5)*	Tony Lazzeri — 1.156 (21)*
Center field	Hack Wilson — 1.269 (6)*	Earl Averill — 1.213 (10)*
Left field	Al Simmons — 1.253 (7)*	Bob Johnson — 1.189 (14)
Catcher	Bill Dickey — 1.179 (16)*	Gabby Hartnett — 1.140 (25)*
Shortstop	Joe Cronin — 1.169 (20)*	Arky Vaughan — 1.086 (36)*
Third base	Pie Traynor — 1.109 (31)*	Pinky Higgins — 1.068 (38)

* Indicates that a player was elected to the Baseball Hall of Fame.

[1] The number after each player's name is his potential runs per game (PRG) rating, and the number in parentheses is his PRG ranking for the LBE.

of .671 and an earned run average of 2.28. Ruth was also a very good hitter. In 1918 the Red Sox used him as both a pitcher and an everyday player and in 1919 used him primarily as an everyday player. Ruth responded with 29 home runs, a new major league record. After the season, the Red Sox owner Harry Frazee sold him to the New York Yankees in order to obtain cash for his foundering theater business. New York's Polo Grounds had a shorter right-field fence than Boston's Fenway Park, and Ruth hit 54 home runs his very first year with the Yankees. The Yankees soared, the Red Sox crashed, and the "curse of the Bambino" was born.

Babe Ruth went on to compile some astounding statistics: he led the American League in slugging average 13 times, home runs 12 times, walks 11 times, on-base percentage 10 times, runs eight times, runs batted in six times, and batting average once. He ranks first all-time in slugging average, on-base plus slugging, and RBI percentage and second all-time in on-base percentage and home run percentage. Ruth holds the record for the most extra-base hits in a season (119 in 1921), is second in the number of seasons with 150 or more runs batted in (five), and is tied for third with two seasons of 400 or more total bases. He held the record for home runs in a season (60) for 34 years until it was broken by Roger Maris (in 1961) and for career home runs (714) for 39 years until it was broken by Hank Aaron (in 1974). Babe Ruth was more than just a great slugger. He had a lifetime batting average of .342 (ninth all-time), a lifetime on-base percentage of .447 (second all-time), 2,062 walks (third all-time), and a walk-to-strikeout ratio of 1.53 (sixth highest all-time among the top 50 all-time home run hitters). Ruth was elected to the Baseball Hall of Fame in 1936, as one of its five charter members.

As had been the case in the Era of Constant Change, three of the four leading hitters in this era were big, slugging first basemen. The highest-ranking first baseman and the second-highest-ranking hitter overall was Lou Gehrig, with a PRG rating of 1.429. Gehrig played from 1923 to 1939: he spent his entire 17-year major league career with the New York Yankees. He was nicknamed the "Iron Horse" because he was so durable. Gehrig held the record for the most consecutive games played (2,130) for 56 years until it was surpassed by Cal Ripken, who ended up playing in an astounding 2,632 consecutive games.

Lou Gehrig played in the shadow of Babe Ruth for most of his career, but he amassed some incredible numbers of his own. He led the American League in runs batted in and on-base percentage five times each; in runs four times; in home runs and walks three times each; in doubles and slugging average twice each; and in hits, triples, and batting average once each. His accomplishments go far beyond these single-season triumphs. Lou Gehrig was actually one of the greatest hitters ever. He ranks third all-time in slugging average and on-base plus slugging and has the all-time record for seasons of 400 or more total bases (five). Gehrig is one of only three players to have two seasons of 100 or more extra-base hits and ranks second all-time in extra-base hits per at bat. He leads all players with seven seasons of 150 or more runs batted in and ranks second all-time in runs batted in per at bat (.249). Gehrig holds the all-time record for grand slam home runs (23).

Lou Gehrig was not a one-dimensional slugger. His lifetime batting average of .340 ranks third all-time for first basemen, and he ranks fifth all-time for all players in on-base percentage. Among the top 50 all-time home run hitters, he has the third-highest walk-to-strikeout ratio (1.91). His fielding average was comparable to the other leading first basemen of his era. Gehrig was surprisingly fast for a first baseman and was second in his era for stolen bases by a first baseman, including 15 steals of home plate.

Gehrig won one Triple Crown, won two Most Valuable Player Awards, and was elected to the Baseball Hall of Fame by a special committee in 1939 right after he retired, instead of having to wait the usual five years after retiring. He had to retire when he was only 35 years old because he contracted a rare muscular disease, amyotrophic lateral sclerosis, ever since referred to as Lou Gehrig's disease. His optimistic farewell words will always be remembered by baseball fans: "I may have had a tough break, but I have an awful lot to live for." Gehrig died less than two years later.

The second-highest-ranking first baseman and the third-highest-ranking hitter overall was Hank Greenberg, with a PRG rating of 1.381. Greenberg played from 1930 to 1947: 12 seasons with the Detroit Tigers and one season with the Pittsburgh Pirates. "Hammering" Hank Greenberg, at 6 feet 3.5 inches and 210 pounds, was the quintessential first baseman slugger. He led the American League in home runs and runs batted in four times each. In 1935 and 1940 he led the league in both events and was awarded Most Valuable Player both times. Greenberg also led the American League in walks and doubles twice each and in runs and slugging average once each.

Greenberg hit over .300 eight years in a row (1933–1940), but in 1946 he was the first player ever with more than 40 home runs and a batting average under .300. The sacrifice of singles for home runs (i.e., the reverse philosophy of a bird in the hand is worth two in the bush) has since become a common occurrence, especially in recent years.[2] Greenberg had 150 or more runs batted in three different times, a feat exceeded by only three other players. He ranks third all-time for the most RBI in a single season (183 in 1937), third in RBI per at bat, and seventh in slugging average and on-base plus slugging. His statistics would have been even better had he not lost one season because of an injury and nearly four and a half seasons because of his military service in World War II. Greenberg was elected to the Baseball Hall of Fame in 1956.

The third-highest-ranking first baseman and the fourth-highest-ranking hitter overall was Jimmie Foxx, with a PRG rating of 1.366. Foxx played from 1925 to 1945: 11 years with the Philadelphia Athletics, six years with the Boston Red Sox, and three years split between the Chicago Cubs and Philadelphia Phillies. He had two nicknames: "Double X," for the spelling of his last name, and "the Beast," because he was so strong and hit such long home runs.

Jimmie Foxx succeeded Babe Ruth as baseball's leading slugger. He hit more home runs in the 1930s (415) than any other player. Foxx led the American League in home runs four times, slugging average five times, runs batted in three times, on-base percentage three times, walks and batting average twice, and runs once. He won a Triple Crown and three Most Valuable Player Awards. Foxx ranks third all-time in number of seasons with 150 or more runs batted in (four) and is tied for third all-time in seasons of 400 or more total bases (two). He has all-time career rankings of fifth in runs batted in percentage and slugging average, sixth in on-base plus slugging, and 10th in on-base percentage. Foxx, like Gehrig and Greenberg, may have had even better statistics had his career not been shortened. His heavy drinking caught up with him and his slugging dropped off significantly in 1941, when he was only 34 years old. He played parts of three more seasons but never recovered his slugging prowess. Foxx was elected to the Baseball Hall of Fame in 1951.

The highest-ranking second baseman and the fifth-highest-ranking hitter overall was Rogers Hornsby, with a PRG rating of 1.283. This was the highest ranking for any second baseman up to that time. Second basemen were valued much more for their fielding ability than for their hitting ability. Scouts were skeptical of Hornsby's hitting ability in the minor leagues, and he was brought up to the major leagues more for his glove than for his bat. Hornsby developed into a great hitter only after changing his batting stance from crouching over and choking up on the bat to standing up straight and holding the bat at the end.

Hornsby played from 1915 to 1937: 12 seasons with the St. Louis Cardinals, one season each with the New York Giants and Boston Braves, and three seasons with the Chicago Cubs. At that point his playing career was essentially over because he did not recover from a broken ankle incurred while sliding into a base. Hornsby continued, nevertheless, as a part-time player for six more years. Probably he hoped to join the 3,000-hit club—at that time there were only six players in that select group—but he managed only 75 more hits for a career total of 2,930.

There are some similarities between Rogers Hornsby and Ty Cobb. They were about the same size (Cobb 6 feet tall and 175 pounds and Hornsby 5 feet 11 inches tall and 175 pounds), and both had high lifetime batting averages (Cobb .366, the highest ever for a left-handed batter and for an American League player, and Hornsby .358, the highest ever for a right-handed batter and for a National League player). Both won Triple Crowns—Cobb one and Hornsby two. They both managed major league teams toward the end of their playing careers—Cobb for six years and Hornsby for 14 years. Perhaps the biggest similarity between them was their nastiness. They were temperamental, violent, and often in trouble. Cobb and Hornsby were great baseball players, but they were, at times, very difficult to get along with.

Hornsby led the National League in hits, doubles, and runs batted in four times each, batting average seven times, on-base percentage eight times, and slugging average nine times. Most of these statistics were compiled in the six-year period from 1920 to 1925, some of the most productive years any player has ever had. He led the league in batting average, on-base percentage, and slugging all six years and in hits, doubles, and runs batted in four times each, winning two Triple Crowns and one Most Valuable Player Award. He later won a second Most Valuable Player Award. His all-time career rankings include second in batting average, eighth in on-base percentage and on-base plus slugging, and 12th in slugging average. Hornsby's single-season accomplishments include one season with more than 100 extra-base hits (12th all-time), one season with more than 150 runs batted in, two seasons with 400 or more total bases (tied for third all-time), and the second-highest total bases season ever (450). Hornsby was elected to the Baseball Hall of Fame in 1942.

The highest-ranking center fielder and the sixth-highest-ranking hitter overall was Hack Wilson, with a PRG rating of 1.269. Wilson played from 1923 to 1934: three mediocre years with the New York Giants, six very productive years with the Chicago Cubs, and three declining years with the Brooklyn Dodgers. He did not look like a center fielder—he was only 5 feet 6 inches tall but weighed 190 pounds. Wilson was big from the waist up but had very short legs and small feet. He was called "Hack" because he looked like George Hackenschmidt, a popular wrestler of that time. Wilson was a drinker and a brawler who once considered a career in professional boxing.

Wilson may not have looked like a center fielder, but he was capable enough to play over 900 games at that position. His greatest asset, however, was his hitting. Wilson led the National League in home runs four times, in walks and runs batted in twice each, and in slugging average once. While with Chicago he had a batting average of .322 and a very high RBI rate (per at bat) of .244. In 1930 Wilson had his best year: he set a single-season major league record for runs batted in (191), which still stands, led the National League with a slugging average of .723, and hit 56 home runs, a National League record that lasted 68 years until it was broken by Mark McGwire and Sammy Sosa in 1998. After the 1930 season, Wilson's statistics declined to ordinary levels, but he ended up with a lifetime batting average of .307. He ranks ninth all-time in runs batted in per at bat (.223) and is one of only nine players with two or more seasons of 150 runs batted in. Wilson was elected to the Baseball Hall of Fame by the Veterans Committee in 1979.

The highest-ranking left fielder and the seventh-highest-ranking hitter overall was Al Simmons, with a PRG rating of 1.253. Simmons played from 1924 to 1944: nine seasons with the Philadelphia Athletics, three seasons with the Chicago White Sox, two seasons with the Washington Senators, and six seasons with three other teams. Simmons's real name was Aloys Syzmanski, but he changed it to Al Simmons. He was called "Bucket Foot" because of his unorthodox batting style. As he stood in the batter's box, his left foot was pointed toward third base and he stepped in that direction as he swung at the ball. This is referred to as stepping into the bucket because, in the old days, the water bucket was on the players' bench in the third base dugout. It appeared awkward, but Simmons was able to hit effectively because his upper body was lined up with the pitcher.

Simmons was a line-drive hitter. He hit 307 home runs, but his batting average and runs batted in were more important. In each of his first 11 years he batted over .300 and had over 100 runs batted in. He batted over .380 four times and once had 253 hits, which ranks fifth highest all-time. Simmons is tied for fourth with three seasons of more than 150 runs batted in.

While with Philadelphia, Simmons batted third and Jimmie Foxx batted fourth, making them a duo rivaling Ruth and Gehrig of the Yankees. When Simmons was traded to Chicago, he no longer had the support of Jimmie Foxx or the hitter-friendly confines of Shibe Park, and his statistics declined. Nevertheless, Simmons ended his career with a lifetime batting average of .334, good for 21st all-time and 8th all-time for a right-handed batter. He was one of the better outfielders of his era, being possessed with good hands and a strong throwing arm. Simmons was elected to the Baseball Hall of Fame in 1953.

The second-highest-ranking right fielder and the eighth-highest-ranking hitter overall was Harry Heilmann, with a PRG rating of 1.233. Heilmann played from 1914 to 1932: 15 years with the Detroit Tigers and two years with the Cincinnati Reds. He was nicknamed "Slug," but he actually hit more for average than for power. At 6 feet 1 inch tall and 195 pounds, he was a slow runner and started out as a first baseman, but he was awkward there and made a lot of errors. When Ty Cobb became manager of the Tigers in 1921, he shifted Heilmann to right field and modified his batting style. As soon as Heilmann stood in a crouch and put more weight on his front foot—almost the opposite of the adjustment that Rogers Hornsby made early in his career—his batting average improved from .309 to .394!

Heilmann had a batting average over .300 12 years in a row (including one year over .400 and three years in the .390s), for a career batting average of .342, the third-highest lifetime batting average ever for a right-handed batter. He led the American League in hits and doubles once each and in batting average four times, curiously all of them in alternate odd-numbered years (1921, 1923, 1925, and 1927). Two of Heilmann's batting titles involved dramatic performances on the last day of the season. Heading into the last day of the 1925 season, Heilmann trailed Tris Speaker by a fraction of a point. Speaker had leg problems and did not play the last day. After the first game of a doubleheader, Heilmann was ahead of Speaker by a fraction of a point. He was urged to sit out the second game and assure himself of the batting title, but he refused. In the second game, Heilmann improved his average by four points and won the batting title .393 to Speaker's .389. Heading into the last day of the 1927 season, Heilmann trailed Al Simmons by a fraction of a point. Simmons played his first game in the Eastern Time Zone and improved his batting average by a fraction of a point more. Heilmann played the first game of a doubleheader in the Central Time Zone and finished with an average three points higher than Simmons. Again urged to sit out the second game, Heilmann refused and finished the day with another batting title, an average of .398 to Simmons's average of .392.

The 1951 All-Star Game was played in Detroit, but Harry Heilmann could not attend because he was suffering in the last stages of lung cancer. Ty Cobb thought he had convinced baseball to have a special election, induct Heilmann into the Hall of Fame, and present him with a bronze plaque at the All-Star Game. Cobb told Heilmann, but Heilmann died the day before the game thinking he had already been elected. Heilmann was elected to the Hall of Fame the following year (1952).

The third-highest-ranking right fielder and the ninth-highest-ranking hitter overall was Mel Ott, with a PRG rating of 1.223. Ott played from 1926 to 1947, spending his entire 22-year major league career with the New York Giants. Mel, short for his first name Melvin, was an appropriate nickname for this small, soft-spoken, all-around player. He was so popular that the right-field stands at the Polo Grounds were known as Ottville, in recognition of the many home runs he hit there and the superb way he played the right-field position. At 5 feet 9 inches and 170 pounds, he had a distinctive batting style. When the ball was on its way to the plate, Ott took a high step forward with his front foot and raised his bat. As the ball got closer he stepped forward with the front foot and lowered his bat to a horizontal position and took a smooth, level swing.

Mel Ott came to the Giants when he was only 17 years old, having never played a game in the minor leagues. After two years of part-time play and close tutelage by manager John McGraw, he became the regular right fielder. Ott was only 19 years old. The following year he hit 42 home runs, drove in 151 runs, and had 113 walks, a batting average of .328, and a slugging average of .635, one of the most successful seasons a 20-year-old has ever had.

Ott's 20-year-old season may have been the best he ever had, but he had many others nearly as good. He led the National League in home runs and walks six times each, in on-base percentage four times, in runs twice, and in runs batted in and slugging average once each. In five consecutive games, Ott was walked four times each—more than anyone else in the entire history of the game. He batted over .300 11 times, for a lifetime batting average of .304. Ott finished his career with 511 home runs, a National League record until Willie Mays came along. He had nine seasons of over 100 runs batted in. Ott was never given a Most Valuable Player Award, but he was selected for 11 consecutive All-Star Games. Mel Ott was elected to the Baseball Hall of Fame in 1951.

The second-highest-ranking center fielder and the 10th-ranked hitter overall was Earl Averill, with a PRG rating of 1.213. Averill played from 1929 to 1941: 10 excellent seasons with the Cleveland Indians and three part-time seasons split among Cleveland, Detroit, and Boston. His batting statistics deteriorated after 1937 when a spinal problem prompted him to adjust his swing.

Averill was known as the "Earl of Snohomish"—he was born and raised in Snohomish, Washington. The only major batting events in which he led the American League were hits and doubles (1936). He was, however, a steady player year in and year out, at one point playing in 673 straight games. Averill hit .300 or more in 8 of his 10 years with Cleveland and finished his career with a batting average of .318. Although only 5 feet 9 inches tall and 172 pounds, he was able to generate enough power to take advantage of the right-field fence at Cleveland League Park. He had 238 career home runs. Averill was chosen for the first six All-Star Games and was elected by the Veterans Committee to the Baseball Hall of Fame in 1975.

The highest-ranking catcher and the 16th-ranked hitter overall was Bill Dickey, with a PRG rating of 1.179. Just plain Bill (he had no nickname) played from 1928 to 1946, spending his entire 17-year major league career with the New York Yankees. Dickey never led the league in a major batting event. He did, however, catch more than 100 games 13 years in a row (the first catcher ever to do so) and batted over .300 in 10 of those seasons. Dickey also had more than 100 runs batted in four times and hit 202 home runs, a lot for a catcher at that time. He was regarded as a good-fielding catcher—he led the league in fielding four times—and was known for his pitch selection and positive influence on his teammates. Dickey was selected for 11 All-Star Games and was elected to the Baseball Hall of Fame in 1954.

The highest-ranking shortstop and the 20th-ranked hitter overall was Joe Cronin, with a PRG rating of 1.169. Cronin played from 1926 to 1945: two part-time years with the Pittsburgh Pirates, seven years as a regular with the Washington Senators, and 11 years with the Boston Red Sox, the last four years of which he was a part-time player. He was the team manager the last two years in Washington and all of his years in Boston.

Cronin led his league in only two offensive events: twice in doubles and once in triples. He did, however, bat over .300 eight times and finished his career with a lifetime batting average of .301. Cronin hit for average more than for power, but he did hit 170 home runs and drive in 1,424 runs, making him the best overall hitting shortstop of his era. He ranked only 20th overall for his era, but shortstops were not valued as much for their bats as for their gloves and arms. Cronin was elected to the Baseball Hall of Fame in 1956 and became the president of the American League in 1959, a position he held until 1974.

The highest-ranking third baseman and the 31st-ranked hitter overall was Pie Traynor, with a PRG rating of 1.109. Traynor played from 1920 to 1937, spending his entire 17-year major league career with the Pittsburgh Pirates, the first two years and last two years of which were as a part-time player. He was a great team player and a nice person—a quiet, polite gentleman, whose nickname "Pie" came from his favorite dessert. Traynor hit for average but not for power— a lifetime batting average of .320 but only 58 home runs. The only major batting event in which he ever led the National League was doubles, and he did that only once. Traynor had 164 career triples, facilitated by his ability to hit line drives in the outfield gaps at spacious Forbes Field. Traynor had 10 or more triples in a season 11 times and hit over .300 10 times in his 13 full seasons. He was a good third baseman—he had a great glove and a strong arm, but his arm was erratic and he led his position at fielding only once. Traynor was elected to the Baseball Hall of Fame by the Veterans Committee in 1948.

A total of 45 position players from this era were elected to the Baseball Hall of Fame. This is more than the total elected for the first three eras of baseball history combined (37) and represents almost one-third of all 143 position players elected in the entire history of Major League Baseball to the present time. Why were so many from this era elected to the Hall of Fame? This change may be explained by the fact that (1) the Hall of Fame wasn't founded until toward the end of this era, and the early elections to the Hall occurred at a time when the exploits of these players were fresh in mind; and (2) most position players are elected to the Hall of Fame based on their hitting accomplishments, and this was a live ball era in which offensive statistics increased dramatically over those of the previous dead ball era.

Sixteen of the 18 leading hitters listed in table 4.2 were elected to the Hall of Fame. Another 16 players not in table 4.2 were also elected to the Hall of Fame based primarily on their hitting. This was not an era known for its great fielders, and a case can be made that only four players were elected based mainly on their fielding: Max Carey, Rabbit Maranville, Joe Tinker, and Lloyd Waner. Both hitting and fielding credentials were most likely factors in the election of nine others: Mickey Cochrane, Earle Combs, Rick Ferrell, Frankie Frisch, Chick Hafey, Billy Herman, Freddie Lindstrom, Edd Roush, and Joe Sewell.

* 5 *

The Live Ball Continued Era, 1942–1962
THE AGE OF TED WILLIAMS

Joe had played for the Yankees, Boston's bitter rival, and the two great hitters—Joe, who hit in 56 straight games in 1941, and Ted, who hit .406 that same year—had had their own unspoken rivalry. They were at the center of one of the great ongoing debates among baseball fans of several generations, as to who the better ball player was, and what might have happened had each played for the other's team, in the other's ballpark. The summary judgment: Joe was the better all-around player; Ted, the better hitter.

David Halberstam, *The Teammates,* p. 17

World War II broke out in Europe in September 1939, but it wasn't until December 1941 that Pearl Harbor was attacked and the United States entered the war. Some major league players immediately enlisted in the armed forces and did not play for the duration of the war. Most of the major league players who later joined the armed forces continued to play baseball in 1942 and then went into the service and missed the 1943, 1944, and 1945 baseball seasons. In the 1942 season there was an across-the-board decline in batting statistics in both major leagues. American League batting statistics were even lower from 1943 to 1945. National League batting statistics were mixed, some a little higher and some a little lower than in 1941. After the war, batting statistics stabilized until 1963, when a downturn occurred, resulting in another new era. The net result of all of these changes is reflected in the summary below.

	AVG	OBP	SLG	OPS	HR%	ERPG
Live Ball Era (1921–1941)	.281	.344	.398	742	1.4	8.26
Live Ball Continued (1942–1962)	.259	.331	.382	712	2.1	7.66
Dead Ball Interval (1963–1976)	.249	.329	.368	697	2.2	7.03

Thus, there is nearly an across-the-board decline in the batting statistics as we move from era, to era, to era. The only exception to this decline is the steady increase in home run percentage. It seems counterintuitive, but home run percentage was actually higher in the Dead Ball Interval than in both phases of the live ball era. We will return to this trend in future chapters.

The top 10 hitters of the Live Ball Continued era for each of the average measures of batting performance are identified in table 5.1. The pattern of agreement and disagreement between the measures is similar to the pattern we have seen in the preceding eras. If you look at the top 10 as a group, there is general agreement on eight players who belong there. All the measures agree on five players (Williams, DiMaggio, Mize, Musial, and Mantle), all except one measure agree on two players (Kiner and Mays), and all except two measures agree on one other player (Keller). Eighteen players share the top 10 rankings—and 10 share the top five rankings.

If you look at the specific ranking of the players within the top 10, there is more disagreement. Ted Williams is ranked number 1 by all measures, the first player to be so ranked since Dan Brouthers in the Era of Constant Change. Stan Musial is ranked number 4 by all except two measures. There is considerable disagreement on the rankings of the other 10 players. None of them have more than three of any specific ranking, and two of them—Gordon and Mathews—are unranked by more measures than they are ranked. The widespread differences in the way the various measures rank the hitters of this era present a problem for the person who wants to identify the best hitters of this era. Which of the measures should one choose, and what criteria should govern that choice? A concise statement pertaining to that choice can be retrieved from the Pregame Analysis.

Table 5.2 summarizes the leading hitters of the Live Ball Continued era by the positions they played in the field. Nine of the 10

TABLE 5.1 The Top 10 Hitters in the LBC Era by Various Measures

Player	PRG	Advanced Weighted Measures			Basic Weighted Measures			Unweighted Measures	
		LSLR	RC/27	LWTS	TA	OPS	SLG	OBP	AVG
Ted Williams	1	1	1	1	1	1	1	1	1
Joe DiMaggio	2	2	3	2	4	3	2	8	3
Johnny Mize	3	8	5	7	7	5	3	10	4
Stan Musial	4	4	4	4	3	4	4	4	2
Charlie Keller	5	—	6	8	5	8	9	5	—
Mickey Mantle	6	3	2	3	2	2	5	3	9
Ralph Kiner	7	4	8	5	6	6	7	8	—
Duke Snider	8	—	10	—	10	9	8	—	—
Sid Gordon	9	10	—	10	—	—	—	—	—
Willie Mays	10	6	7	6	8	7	5	—	6
Eddie Mathews	(20)	7	—	—	—	10	10	—	—
Jackie Robinson	(25)	—	9	9	9	—	—	6	5
Other players[1]	—	—	—	—	—	—	—	—	—

[1] Six other players ranked in the top 10 but in one measure only: Ferris Fain #2 in OBP, Roy Cullenbine #7 in OBP, Enos Slaughter #7 in AVG, Al Rosen #9 in LSLR, Carl Furillo #8 in AVG, and Ted Kluszewski tied for #9 in AVG.

leading hitters were outfielders, and five of those nine were left fielders. The highest-ranking left fielder and the highest-ranking hitter overall was Ted Williams, with a PRG rating of 1.428. Williams played from 1939 to 1960, with two breaks for military service in World War II and the Korean War. He spent his entire 19-year major league career with the Boston Red Sox. He had several nicknames: "The Kid" because of his youthful look, the "Splendid Splinter" because of the contrast between his hitting ability and thin physique, the "Thumper" because of his great hitting ability, and "Teddy Ballgame" because of his lifetime absorption in the science of hitting a baseball. Among the top 50 all-time home run hitters, he has the highest walk-to-strikeout ratio of 2.85.

Williams was the greatest player ever in getting on base. He holds the record for the number of consecutive games in getting on base— 84 in 1949. Williams also ranks third and fourth in that statistic

TABLE 5.2 Leading Hitters of the LBC Era by Position[1]

Position	Leading Hitter	Second-Leading Hitter and Others in the Top 10 Overall
Left field	Ted Williams — 1.428 (1)*	Stan Musial — 1.234 (4)*; Charlie Keller — 1.230 (5); Ralph Kiner — 1.203 (7)*; Sid Gordon — 1.174 (9)
Center field	Joe DiMaggio — 1.326 (2)*	Mickey Mantle — 1.214 (6)*; Duke Snider — 1.178 (8)*; Willie Mays — 1.169 (10)*
First base	Johnny Mize — 1.265 (3)*	Rudy York — 1.154 (13)
Third base	Al Rosen — 1.164 (11)	Eddie Mathews — 1.113 (20)*
Catcher	Roy Campanella —1.159 (12)*	Yogi Berra — 1.121 (18)*
Right field	Tommy Henrich — 1.138 (16)	Vic Wertz — 1.134 (17)
Second base	Jackie Robinson — 1.104 (24)*; Bobby Doerr — 1.104 (24)*	
Shortstop	Ernie Banks — 1.074 (36)*	Lou Boudreau — 1.011 (47)*

* Indicates that a player was elected to the Baseball Hall of Fame.
[1] The number after each player's name is his potential runs per game (PRG) rating, and the number in parentheses is his PRG ranking for the LBC era.

with 69 in 1941 and 65 in 1948. He led the American League in on-base percentage 12 times, the all-time major league record. His career on-base percentage of .482 is the highest of any player who ever played the game.

Ted Williams was one of the greatest Triple Crown–type hitters of all time (see chap. 13). Only he and Rogers Hornsby have two Triple Crowns, and Williams missed a third by the narrowest of margins (a batting average of .3428 to George Kell's .3429 in 1949). He even won a Triple Crown in the minor leagues. Williams also led the league in batting average and home runs one other time, which makes him the only player in major league history to lead his league in batting average and home runs three times. He is also one of only five players to lead both leagues in each Triple Crown event (1942).

Williams is the last player to have a batting average over .400 (.406 in 1941), a mark that has lasted 68 years. He led the American League in batting average six times, the first time the year he turned 23 and the last time the year he turned 40. Williams also led the American League in slugging average and walks eight times each, runs six times, home runs and runs batted in four times each, and doubles twice. His lifetime batting average of .344 ranks seventh all-time. He ranks second all-time in slugging average and on-base plus slugging and fourth all-time in runs batted in percentage (.239).

Ted Williams was a controversial player—outspoken and often at odds with the press and Red Sox fans. Early in his career he vowed never again to tip his hat to the crowd, a vow that he kept until long after he hit a home run in his last major league at bat. It wasn't until Ted Williams Day in 1991 that he finally tipped his hat.

Williams's batting statistics are even more amazing when consideration is given to two additional facts. He lost three years of playing time because of his military service in World War II and two additional years because of his military service in Korea and injuries. Williams missed the years he turned 25, 26, and 27, prime years for most players, and the years when he turned 34 and 35, compelling him to play into his forties in order to establish his place in baseball history.

The second element that makes his batting statistics even more amazing is the "Boudreau Shift." Beginning in 1946, Cleveland manager Lou Boudreau placed six of his fielders on the right side of second base, making it more difficult for Williams to pull the ball. The shift was adopted by other managers, and Williams was widely criticized for trying to hit through the shift instead of taking the easy safe hit to the left side of the field. His batting average did decline after employment of the shift, but part of that decline can be attributed to the aging process. Ted Williams was elected to the Baseball Hall of Fame in 1966 and in his induction ceremony speech pled for the election of Negro League baseball players to the Hall of Fame. A Negro Leagues Committee was subsequently established, and several Negro League players were elected to the Hall of Fame in the 1970s and many more thereafter.

The highest-ranking center fielder and the second-highest-ranking hitter overall was Joe DiMaggio, with a PRG rating of 1.326. DiMaggio played from 1936 to 1951 (except for three years during

World War II) and spent his entire 13-year career with the New York Yankees. He had two nicknames: the "Yankee Clipper" and "Joltin' Joe"—the former because he moved with such effortless grace, and the latter because he was also such a great slugger. As David Halberstam noted above, DiMaggio was much more than just a graceful slugger. He had a lifetime batting average of .325. DiMaggio had an exceptionally good batting eye—his at bats per strikeout ratio of 18.5 was one of the highest ever for a home run hitter. He was a great fielder and a great base runner—his manager, Joe McCarthy, said he was the best base runner he had ever seen.

DiMaggio's most productive years were before World War II. In seven seasons he had a combined batting average of .340, an on-base percentage of .522, slugging average of .607, and on-base plus slugging of 1129. DiMaggio received three Most Valuable Player Awards and in 1941 set one of the most revered and enduring records in the history of baseball—hitting safely in 56 consecutive games. That same year, he was on base in 74 consecutive games, the second highest on-base streak ever.

DiMaggio missed three years because of World War II, and they proved to be very costly years. He was 32 years old when he returned to baseball, and several injuries limited his postwar productivity. However, he continued to be chosen to play in the annual All-Star Game. DiMaggio ended up playing in the All-Star Game in every year of his 13-year major league career. He led the American League in home runs, runs batted in, batting average, and slugging average twice each and in runs and triples once each. DiMaggio ranks eighth all-time in runs batted in percentage, 10th all-time in slugging average, and 13th all-time in on-base plus slugging. Joe DiMaggio was elected to the Baseball Hall of Fame in 1955.

The highest-ranking first baseman and the third-highest-ranking hitter overall was Johnny Mize, with a PRG rating of 1.265. Mize played from 1936 to 1953: six years with the St. Louis Cardinals, five years with the New York Giants, and four years with the New York Yankees. He had six excellent minor league seasons but was not brought up to the major leagues until he was 23 years old because of a recurring hip problem. Mize also lost three years in the major leagues because of his service during World War II. He had good batting statistics with the Cardinals and for the first four years with the Giants. When his batting statistics declined, he was traded to the Yankees and suffered a shoulder separation, causing him to be used primarily as a pinch hitter and occasional starter.

Johnny Mize was a big man—6 feet 2 inches tall and 215 pounds. He was primarily a slugger, but he also hit well for average and was a good-fielding first baseman. Mize was called the "Big Cat" because he was very good at fielding bad hops at first base and because he had such a graceful stance at the plate. Although he was primarily a straightaway hitter, Mize learned to pull the ball when he was traded to the Giants and had to play half of his games at the lopsided Polo Grounds. He led the National League in home runs and slugging average four times each, in runs batted in three times, and in runs, doubles, triples, and batting average once each.

Mize has all-time career rankings of 17th in runs batted in percentage and slugging average and 21st in on-base plus slugging. He had a lifetime batting average of .312. Mize is the only player in Major League Baseball history with more than 50 home runs and fewer than 50 strikeouts in a single season—in 1947 he hit 51 home runs but struck out only 42 times. His 51 home runs constituted a National League record for a left-handed batter until it was broken in 2001 by Barry Bonds. Mize was selected to play in ten All-Star Games and was elected to the Baseball Hall of Fame by the Veterans Committee in 1981.

The second-highest-ranking left fielder and the fourth-highest-ranking player overall was Stan Musial, with a PRG rating of 1.234. Musial played from 1941 to 1963 and spent his entire 22-year major league career with the St. Louis Cardinals. His nickname, "Stan the Man," stems from a series of games when Brooklyn Dodger fans chanted "here comes the man, here comes the man" whenever Musial came to the plate.

Stan Musial was a quiet and unassuming man, but his bat spoke volumes. He was one of the greatest players of his time, and there were four all-time greats he had to compete with—Joe DiMaggio in the 1940s, Ted Williams in the 1940s and 1950s, and Mickey

Mantle and Willie Mays in the 1950s and early 1960s. Musial was a balanced high-percentage and long ball hitter. He led the National League in doubles eight times; batting average seven times; hits, on-base percentage, and slugging average six times each; runs and triples five times each; runs batted in twice; and walks once. Musial did not strike out often: among the 50 all-time home run leaders, he ranks highest in at bats per strikeout (15.8) and has the second-highest walk-to-strikeout ratio (2.30). His best year was 1948, in which he led the National League in eight batting events; his 429 total bases that year ranks sixth highest all-time and his 103 extra-base hits ranks seventh highest all-time for a single season. Musial ranks 14th all-time in on-base plus slugging and 21st in slugging average. He received three Most Valuable Player Awards and played in 19 All-Star Games. Musial was elected to the Baseball Hall of Fame in 1969.

The third-highest-ranking left fielder and the fifth-highest-ranking hitter overall was Charlie Keller, with a PRG rating of 1.230. Keller played from 1939 to 1952, spending 11 seasons with the New York Yankees and two seasons with the Detroit Tigers. He detested the nickname "King Kong" given to him because he was a strong muscular man with black hair on his body.

Charlie Keller might have been one of the very best power hitters of all time, but he was a full-time major league player for only six years. The Yankees delayed two years before bringing him up to the majors, and a congenital back disorder prematurely reduced his role to that of a part-time player. The only batting event in which Keller ever led the league was walks (twice); he had over 100 walks in five of his six full-time seasons. His RBI percentage of 20.1 (24th all-time) helps explain his high PRG rating. Keller was selected to play in five All-Star Games but has never been elected to the Baseball Hall of Fame.

The second-highest-ranking center fielder and the sixth-highest-ranking player overall was Mickey Mantle, with a PRG rating of 1.214. Mantle played from 1951 to 1968, spending his entire 18-year major league career with the New York Yankees. In the minor leagues Mantle played shortstop, and in his first year with the Yankees he played right field. The next year Joe DiMaggio retired and

Mantle switched to center field, where he played until the last two years of his career when, hobbled by injuries, he was transferred to first base.

Mickey was not a nickname but his real first name. His father named him after Mickey Cochrane, the great catcher of the Live Ball Era. Mickey's nickname, the "Commerce Comet," came from his days as a football and baseball player at Commerce (Oklahoma) High School. Mantle's father taught young Mickey how to be a switch-hitter, which proved to be so successful that Mickey Mantle is regarded by many as the greatest switch-hitter in the history of Major League Baseball.

Mantle hit more for power than for average. He led the American League in runs six times, walks five times, home runs and slugging average four times, on-base percentage three times, and doubles, runs batted in, and batting average once each. Not only did Mantle hit a lot of home runs, but he specialized in "monster" or "tape measure" home runs. Early in his career he hit a 565-foot home run over the left-center-field stands at Griffith Stadium in Washington, D.C. Mantle was powerful from both sides of the plate. In every American League park he played in he hit at least one 450-foot home run to both the left and right sides of the field.

Mickey Mantle ranks 14th all-time in home run percentage, 12th all-time in on-base plus slugging, 19th all-time in on-base percentage, but not even in the top 50 in RBI percentage. This helps explain why he ranks only sixth in this era in PRG. Being a feared hitter, Mantle walked a lot, but he also struck out a lot. Among the top 50 home run hitters of all time, Mantle ranks 37th in at bats per strikeout (4.7). He led the league in walks and strikeouts five times and finished his career with 1,733 walks and 1,710 strikeouts. He won one Triple Crown, three Most Valuable Player Awards, and was selected for 16 All-Star Games. Mickey Mantle was elected to the Baseball Hall of Fame in 1974.

The fourth-highest-ranking left fielder and the seventh-highest-ranking hitter overall was Ralph Kiner, with a PRG rating of 1.203. Kiner played from 1946 to 1955: seven years with the Pittsburgh Pirates, two years with the Chicago Cubs, and one year with the Cleveland Indians. He was a right-handed pull-hitting power hitter who benefited from the short left-field fence in Forbes Field. The fence was shortened in 1947 when Hank Greenberg came to Pittsburgh to play one last season in the major leagues. "Greenberg Garden" soon became "Kiner's Corner," as Kiner led the National League in home runs (or tied for the most home runs) in each of his seven seasons with Pittsburgh, for an average of 42 home runs per season.

Kiner hit over .300 three times when he was with Pittsburgh, but he probably would have had a higher batting average if he had adjusted his swing against the teams that shifted players to the left side of the field when he came to bat. On the other hand, his home run production probably would have declined. In addition to home runs, Kiner led the league in slugging average and walks three times each and in runs batted in, on-base percentage, and runs once each. After he was traded to the Indians, his sciatic nerve bothered him and he was able to play only part-time, thus reducing his productivity. He ranks fifth all-time in home run percentage. Kiner was selected to play in six All-Star Games and was elected to the Baseball Hall of Fame in 1975.

The third-highest-ranking center fielder and the eighth-highest-ranking player overall was Duke Snider, with a PRG rating of 1.178. Snider played from 1947 to 1964: 16 years with the Brooklyn / Los Angeles Dodgers and two years with the rival New York / San Francisco Giants. Snider's first name was Edwin. He got his nickname "the Duke" from his father, who thought that his young son behaved like royalty. When Snider grew up, his superior attitude manifested itself in temper tantrums. Snider once claimed that he played baseball for money rather than for fun and would be just as happy if he left the game. Yet when they razed Ebbetts Field, the Duke confessed they also tore a piece from him.

Snider hit more for power than for average. His best years were when the Dodgers were in Brooklyn. Snider led the National League in runs three times, slugging average twice, and hits, walks, home runs, runs batted in, and on-base percentage once each. At one point he hit 40 or more home runs five years in a row. Snider had to compete with both Mickey Mantle and Willie Mays, who played center field for the cross-town Yankees and Giants. The reputations of this trio were enhanced in the "Willie, Mickey, and the Duke" refrain of Terry Cashman's popular song "Talkin' Baseball."

From 1951 to 1959 Snider had the most home runs and RBI in the league, despite the fact that his statistics had begun to decline in 1958 when the Dodgers moved to Los Angeles. The right-field fence in Los Angeles was farther from home plate, and he was plagued by injuries, first a bad knee and then a broken elbow. Although Snider never won a Most Valuable Player Award, he was selected to play in eight All-Star Games and was elected to the Baseball Hall of Fame in 1980.

The fifth-highest-ranking left fielder and the ninth-highest-ranking hitter overall was Sid Gordon, with a PRG rating of 1.174. Gordon played from 1941 to 1955: seven seasons with the New York Giants, four seasons with the Boston / Milwaukee Braves, and a season each with the Pittsburgh Pirates and New York Giants, his original team. He was a power-hitting left fielder who sometimes played third base.

In his first five seasons, Gordon was an average player. In the next five seasons, however, his batting statistics improved—a combined batting average of .293 and averaging 27 home runs and 97 runs batted in per season. Gordon never led his league in any major batting event and was selected to play in two All-Star Games. Sid Gordon has never been elected to the Baseball Hall of Fame.

The fourth-highest-ranking center fielder and the 10th-ranked hitter overall was Willie Mays, with a PRG rating of 1.169. Mays played from 1951 to 1973, spending 20 seasons with the New York / San Francisco Giants and two seasons with the New York Mets. Willie was not a nickname, but his real first name. He was an average-sized (5 feet 11 inches and 180 pounds) but powerful and fast all-around player. Mays was a great outfielder with a great arm, a great base runner, and a great slugger.

Mays holds the all-time record for putouts for an outfielder (7,095) and made what many believe to be the greatest outfield catch and throw ever in the first game of the 1954 World Series. With two runners on base, he raced back to deep center field and caught Vic

Wertz's long drive over his shoulder and immediately spun around and made a perfect throw to the infield, allowing only one runner to tag up and advance to the next base. Mays stole 338 bases in his career, second only to Barry Bonds among the top 50 home run hitters of all time.

Willie Mays led the National League in many different batting events: slugging average five times, home runs four times, triples three times, on-base percentage and runs twice, and hits and walks once each. His all-time rankings are very high in several batting events: third in total bases, fourth in home runs, fifth in extra-base hits, seventh in runs, and 10th in hits. Mays was National League Rookie of the Year, won two Most Valuable Player Awards and 12 Gold Glove Awards, and was selected to play in the All-Star Game 20 years in a row. Willie Mays was elected to the Baseball Hall of Fame in 1979.

The highest-ranking third baseman and the 11th-ranked hitter overall was Al Rosen, with a PRG rating of 1.164. Rosen played from 1947 to 1956, spending his entire 10-year major league career with the Cleveland Indians. During his first three seasons, Rosen had to sit on the bench behind the popular, smooth-fielding Ken Keltner because he had the reputation of being a poor fielder. Rosen demonstrated his hitting ability in his first year as a regular and improved his fielding ability as his short but successful major league career continued.

Al Rosen and Eddie Mathews were the dominant power-hitting third basemen of the era. Rosen had played softball and had been an amateur boxer before joining the Indians. He was nicknamed "Flip" for his style of softball pitching and had his nose broken repeatedly in various boxing matches. Rosen led the American League in home runs and runs batted in twice each and in runs and slugging average once each. His best year was 1953, when he led the American League in runs, home runs, runs batted in, and slugging average. He narrowly missed winning the Triple Crown, as he failed to get a hit in his last at bat. Rosen finished with a batting average of .336 versus .337 for Mickey Vernon. In one five-year stretch, he averaged 165 hits, 31 home runs, and 114 runs batted in. After the 1954 season,

Rosen suffered from whiplash incurred in an automobile accident. The injury bothered him the following year, and as his batting statistics declined, he was widely booed by the Indians fans. After the 1956 season, he abruptly quit the team—he was only 32 years old. Rosen was selected to play in four All-Star Games and won one Most Valuable Player Award but has never been elected to the Baseball Hall of Fame.

The highest-ranking catcher and the 12th-ranked hitter overall was Roy Campanella, with a PRG rating of 1.159. Campanella played from 1948 to 1957, spending his entire 10-year major league career with the Brooklyn Dodgers. He wasn't brought up to the major leagues until he was 27 years old. Prior to that he had spent seven years in the Negro Leagues, where he was one of the most highly regarded players, two successful years in the Mexican League, and two years on Dodgers' minor league teams where he won two Most Valuable Player Awards. Clearly he could have been brought up to the major leagues much sooner.

If it had not been for the times in which he lived, Roy Campanella's major league record probably would have been even more impressive than it was. In his short 10-year major league career he caught for a team of all-stars that won five pennants and lost two others on the last day of the season. Campanella caught more than 100 games nine years in a row, for an average of 126 per year, and averaged 26 home runs and 90 runs batted in for those years. His best year was 1953, when he had a batting average of .312 and a slugging average of .611, hit 41 home runs, and led the National League with 142 runs batted in. Toward the end of his major league career he was bothered by knee injuries and his statistics declined. After his last season he had a bad automobile accident that left him a quadriplegic. The next year Campanella was honored at an exhibition game in the Los Angeles Coliseum before the largest major league crowd ever (90,000+). He won three Most Valuable Player Awards, a feat equaled by only two other National League players before Barry Bonds, and played in eight All-Star Games. Roy Campanella was elected to the Baseball Hall of Fame in 1969.

The highest-ranking right fielder and the 16th-ranked hitter overall was Tommy Henrich, with a PRG rating of 1.138. Henrich played from 1937 to 1950, spending his entire 11-year major league career with the New York Yankees. He was called "Old Reliable" and "the Clutch" for his steady play and timely hitting.

The only major batting events in which he led the American League were triples twice and runs once. You could, however, as his name implies, rely on Henrich—he was a good clutch hitter, steady fielder, and smart all-around player. He played in the shadow of Joe DiMaggio, but he was a vital cog in the Yankee machine. In his first six years, the Yankees won five pennants and four World Series. Henrich played in the All-Star Game his fourth year and again later on in his last four years in the majors. Henrich has never been elected to the Baseball Hall of Fame.

Two players were tied for the highest-ranking second baseman: Jackie Robinson and Bobby Doerr, both of whom ranked 24th overall, with identical PRG ratings of 1.104. Robinson played from 1947 to 1956, spending his entire 10-year major league career with the Brooklyn Dodgers. He was a great all-around athlete. Robinson played four sports at the college level: baseball, football, basketball, and track. He is widely acclaimed as the first black player to play in the modern major leagues, a feat that he accomplished remarkably well despite the pressure he put on himself and the hostility of players and fans.

The only major batting events in which Robinson led the National League were batting average and on-base percentage once each, but he accomplished much more on the playing field. He batted over .300 six straight seasons and finished his career with an average of .311 and an on-base percentage of .409. In 1949 Robinson batted .349, drove in 124 runs, and won the Most Valuable Player Award. He was selected to play in six All-Star Games. Robinson was a very good fielder—he led National League second basemen in fielding three times and in double plays four times. He was also a good base runner—a serious distraction for pitchers concerned that he might steal a base; he led the league in stolen bases twice and stole home plate 19 times.

There are many similarities in the careers of Jackie Robinson and Roy Campanella. Both were black teammates on the Brooklyn Dodgers team that integrated modern baseball. They were both old rookies (Campanella 27 and Robinson 28) and spent their entire 10-year major league careers with only one team. They were both bothered by knee problems that caused their batting statistics to decline in their last two years. Robinson was elected to the Baseball Hall of Fame in 1962.

Bobby Doerr is not as well known to today's fans as Jackie Robinson. He has recently received some notoriety in David Halberstam's book *The Teammates*, a nostalgic account of the lasting friendship of Doerr and three of his Boston Red Sox teammates—Ted Williams, Dom DiMaggio, and Johnny Pesky. Doerr played from 1937 to 1951, spending his entire 14-year major league career as the Red Sox second baseman. He never played a single game at any other position or for any other team.

Early in his career Doerr hit for average but not for power. Later he learned to pull the ball toward the Green Monster and became a slugger. For one 10-year stretch he had a slugging average of .490 and averaged 19 home runs and 101 runs batted in. Doerr had six seasons of 100 or more runs batted in and five seasons with a slugging average of .497 or more. He led the league in slugging average and triples once each. Doerr was a very good fielding second baseman, specializing in double plays and errorless chances. He retired at 33 because of back problems. Doerr was selected to play in nine All-Star Games and was elected to the Baseball Hall of Fame by the Veterans Committee in 1986.

The highest-ranking shortstop and the 36th-ranked hitter overall was Ernie Banks, with a PRG rating of 1.074. Banks played from 1953 to 1971, spending his entire 19-year major league career with the Chicago Cubs. He was voted the "Greatest Cub Player of All Time" and was known as "Mr. Cub." Banks was able to maintain his enthusiastic and positive attitude despite playing with weak teammates on a team that consistently finished in the second division and that only once, late in Banks's career, managed to end up as high as second in the National League standings.

In the second half of his career, Banks was transferred to first base and ended up playing more games at first base than at shortstop (1,259 to 1,125), but he is treated here as a shortstop because his prime years were at that position. Ernie Banks was the first power-hitting shortstop in the major leagues. He led the National League in home runs and runs batted in twice each and in slugging average once. Banks had 40 or more home runs five times, including four seasons in a row, and more than 100 runs batted in eight times, including the same four seasons in a row. He has the second most career home runs per season for a shortstop (37). Alex Rodriguez averaged 43 home runs per season as a shortstop but has since been moved to third base and will likely end his career with many more games as a third baseman, thus moving Ernie Banks back into first place in this statistic for shortstops.

Banks was a durable player—he once played in 717 consecutive games. He was also a good shortstop, leading all National League shortstops in fielding average three times. Banks was selected to play in 11 All-Star Games, won the Most Valuable Player Award two years in a row, and was elected to the Baseball Hall of Fame in 1977 as a first baseman.

A total of 24 position players from this era were elected to the Baseball Hall of Fame. This is about half of the 46 elected in the preceding era. Fourteen of the 15 leading hitters of this era listed in table 5.2 were probably elected based mainly on their hitting. Lou Boudreau was also a good-hitting shortstop, but fielding had to have been a factor because he was outstanding in the field. Three of the nine players not in this table—Nellie Fox, Phil Rizzuto, and Red Schoendienst—were very good fielders and were probably elected for that reason. Richie Ashburn was probably elected based on his all-around abilities because he was a good hitter, fielder, and base runner. George Kell was very good both at the plate and in the field. Joe Gordon was probably elected based on his all-around abilities. Yankee manager Joe McCarthy said Gordon was the best all-around player he ever saw. Pee Wee Reese was probably elected based mainly on his hitting. Larry Doby must have been elected based on his slugging and Enos Slaughter based on his exceptional base running.

6

The Dead Ball Interval, 1963–1976
THE AGE OF HANK AARON

I could never be just another major league player. I was a black player, and that meant I would be separate most of the time from most of the players on the team. It meant I'd better be good, or I'd be gone. . . . If there's a single reason why the black players of the '50s and '60s were so much better than the white players in the National League, I believe it's because we had to be. And we knew we had to be.

Hank Aaron as quoted in *Baseball: The Biographical Encyclopedia*, p. 2

In 1963 Major League Baseball, attempting to redress a perceived imbalance in the game favoring batters over pitchers, expanded the strike zone. The intended effect occurred immediately—batting statistics declined across the board in both leagues. There was, however, also an unintended effect—a decline in attendance. Apparently the fans preferred the old imbalance in favor of batters over pitchers. In 1969 baseball reverted to its previous definition of the strike zone and lowered the maximum height of the pitcher's mound from 15 to 10 inches. Once again the desired on-the-field effect occurred immediately—1969 batting statistics increased across the board in both leagues. And as hoped, the change was accompanied by an increase in attendance. Batting statistics and attendance remained higher from 1969 to 1976 than they had been from 1963 to 1968. The net statistical results for the era as a whole compared with the preceding and succeeding live ball eras are as follows (both leagues combined):

	AVG	OBP	SLG	OPS	HR%	ERPG
Live Ball Continued (1942–1962)	.259	.331	.382	712	2.1	7.66
Dead Ball Interval (1963–1976)	.249	.329	.368	697	2.2	7.03
Live Ball Revived (1977–1992)	.259	.324	.387	711	2.4	7.68

As might be expected, the batting statistics for most of these measures are lower in the Dead Ball Interval than in the two live ball eras. The only exceptions are HR% and OBP. The fact that HR% is higher in the Dead Ball Interval than in the preceding Live Ball Continued era reflects the relentless march of the home run, era by era, from the days of the Dead Ball Era to the present era. The fact that OBP is higher (but only a little higher) in the Dead Ball Interval than in the succeeding Live Ball Revived era probably reflects a more conservative baseball philosophy emphasizing the first step in scoring runs, that is, to get on base.

The top 10 hitters of the Dead Ball Interval for each of the average measures of batting performance are identified in table 6.1. This era has the most disagreement among measures of all the historical eras considered so far. If you look at the top 10 players as a group, there is general agreement on only five players who belong in that group. Two players (Aaron and Robinson) are ranked in the top 10 by all measures, two other players (Allen and Killebrew) are ranked in the top 10 by all except one measure, and one other player (Willie McCovey) is ranked in the top 10 by all but two measures. Twenty-two players share the top 10 rankings, and 14 players share the top five rankings, the most for any era considered so far.

If you look at the specific rankings of the players within the top 10, there is a lot of disagreement. Aaron and Robinson are ranked first or second by all but two measures. Five measures agree that Dick Allen should be ranked third and that Harmon Killebrew should be ranked sixth. Otherwise, not many measures agree on the specific ranking of players in this era.

The extent of the disagreement between the measures in this era can be reduced if we exclude two players from consideration. Rod Carew was a leadoff batter, and Joe Morgan was a number 2 batter. They will be considered in chapter 10. We are left here with the

TABLE 6.1 The Top 10 Hitters in the DBI by Various Measures

Player	PRG	Advanced Weighted Measures			Basic Weighted Measures			Unweighted Measures	
		LSLR	RC/27	LWTS	TA	OPS	SLG	OBP	AVG
Hank Aaron	1	1	2	1	2	1	1	9	3
Frank Robinson	2	2	1	1	1	2	2	3	10
Willie Stargell	3	—	5	7	7	4	4	—	—
Dick Allen	4	3	3	4	3	3	3	6	—
Willie McCovey	5	—	4	3	5	4	5	9	—
Harmon Killebrew	6	6	7	5	6	6	6	7	—
Rico Carty	7	—	—	—	—	—	—	—	5
Orlando Cepeda	8	8	—	—	—	—	7	—	8
Frank Howard	9	—	—	—	—	—	7	—	—
Norm Cash	10	—	8	—	8	7	—	9	—
Al Kaline	(11)	10	9	8	—	8	—	7	7
Carl Yastrzemski	(14)	7	—	10	—	—	—	5	—
Billy Williams	(15)	4	10	9	—	10	9	—	—
Reggie Smith	(16)	—	—	—	10	8	10	—	—
Gene Tenace	(22)	—	—	—	9	—	—	4	—
Joe Morgan	(28)	5	5	6	4	—	—	2	—
Rod Carew	(30)	9	—	—	—	—	—	1	1
Other players[1]	—	—	—	—	—	—	—	—	—

[1] Five other players ranked in the top 10 but in batting average only: Roberto Clemente #2, Tony Oliva #4, Joe Torre #6, Manny Sanguillen #9, and Bob Watson tied for #10.

consideration of only middle-of-the-order batters. The rankings of all of the measures except SLG, OPS, and PRG change and reduce the differences between measures, but significant differences remain. If you want to identify the leading hitters of this era, you still have to decide among the various measures. How you decide which to choose should be governed by the guidance provided in the Pregame Analysis.

Table 6.2 summarizes the leading hitters of the Dead Ball Interval by the positions they played in the field. The two leading hitters were right fielders—Hank Aaron and Frank Robinson. Aaron had a PRG rating of 1.178. He played from 1954 to 1976 with the NL Milwaukee / Atlanta Braves and two years with the AL Milwaukee Brewers. Aaron's 23-year major league career was one of the longest ever. He ranks third all-time in games and plate appearances and second in at bats. Aaron was not a big player—6 feet tall and 180 pounds—but he hit the ball extremely hard, earning himself the nickname of "the Hammer." Unlike most hitters, he hit off his front foot. Aaron was a great student of pitchers and waited patiently for their mistakes. He was not a flashy player—he did play aggressively, but he was quiet about it. Aaron led the National League in doubles, home runs, runs batted in, and slugging average four times each, in runs three times, and in hits and batting average twice each.

Hank Aaron never hit more than 47 home runs in one season, but he held the all-time record for home runs (755) for 33 years, until it was broken by Barry Bonds in 2007. He never drove in more than 132 runs in a season but holds the all-time record of 2,297 runs batted in. Aaron never had more than 92 extra-base hits in one season, but he has the all-time record of 1,477. He never scored more than 121 runs in one season but ranks fourth all-time with 2,174. Aaron never had more than 223 hits in one season but ranks third all-time with 3,378. He never won a Triple Crown, although he came close in 1963 when he led the National League in home runs and runs batted in and came in third in batting average, only seven points behind the leader. Aaron led the National League in home runs and RBI in the same season two other times. Aaron won only one Most Valuable Player Award, but he was selected to play in 21 All-Star Games and was elected to the Baseball Hall of Fame in 1982.

Frank Robinson was the second-leading hitter of the era, with a PRG rating of 1.170. He played from 1956 to 1976: a 21-year major league career including 10 years with the Cincinnati Reds, six years with the Baltimore Orioles, a year with the Los Angeles Dodgers, two years with the California Angels, and two years with the Cleveland Indians. Robinson was an intense and at times fierce competitor. He employed a fearless batting stance with his head and torso extended out over the plate. Robinson was willing to sacrifice his body—he led his league in hit by pitch seven times, for a total of 198

TABLE 6.2 Leading Hitters of the DBI by Position[1]

Position	Leading Hitter	Second-Leading Hitter and Others in the Top 10 Overall
Right field	Hank Aaron — 1.178 (1)*	Frank Robinson — 1.170 (2)*
Left field	Willie Stargell — 1.166 (3)*	Rico Carty — 1.116 (7); Frank Howard — 1.088 (9)
First base	Dick Allen — 1.156 (4)	Willie McCovey — 1.151 (5)*; Harmon Killebrew — 1.148 (6)*; Orlando Cepeda — 1.100 (8)*; Norm Cash — 1.086 (10)
Catcher	Johnny Bench — 1.084 (11)*	Gene Tenace — 1.055 (22)
Third base	Ron Santo — 1.073 (17)	Richie Hebner — 1.011 (28)
Center field	Amos Otis — 1.048 (25)	Jimmy Wynn — 1.015 (27)
Second base	Joe Morgan — 1.011 (28)*	Davey Johnson — .944 (36)
Shortstop	Rico Petrocelli — .973 (35)	Jim Fregosi — .901 (40)

* Indicates that a player was elected to the Baseball Hall of Fame.
[1] The number after each player's name is his potential runs per game (PRG) rating, and the number in parentheses is his PRG ranking for the DBI.

times in his career (third highest ever)—but nevertheless was able to compile impressive batting statistics.

Robinson was primarily a slugger—586 home runs, 1,812 runs batted in, and a slugging average of .537—but he also had a respectable career batting average of .294 and a career on-base percentage of .389. He had one great year at Cincinnati (1962), leading the National League in runs, doubles, on-base percentage, and slugging average, and one great year at Baltimore (1966), leading the American League in runs, on-base percentage, and slugging average and winning the Triple Crown. At Cincinnati he was chosen Rookie of the Year and won one Most Valuable Player Award, and at Baltimore he won one Triple Crown and one Most Valuable Player Award. Robinson is the only major leaguer to win a Most Valuable Player Award in both leagues. He was chosen to play in 12 All-Star Games, six while with Cincinnati and six while with Baltimore. Robinson was the first player to hit a home run out of Baltimore Memorial Stadium and the first Baltimore player to have his jersey (#20) retired. After leaving Baltimore, he became the player-manager of the Cleveland Indians, making him the first black major league manager. He went on to manage several other teams and was chosen American League Manager of the Year several years later. Frank Robinson was elected to the Baseball Hall of Fame in 1982.

The highest-ranking left fielder and the third-highest-ranking hitter overall was Willie Stargell, with a PRG rating of 1.166. Stargell played from 1962 to 1982, spending his entire 21-year major league career with the Pittsburgh Pirates. His nickname "Willie" was a popular version of his real first name, Wilver. Stargell was also affectionately known as "Pops" because he became a father figure for his teammates after the tragic death of Roberto Clemente. Players brought their problems to Pops, and he provided the advice and leadership that enabled his team to maintain their winning ways.

Willie Stargell hit for power rather than for average. He had 475 career home runs but a batting average of only .282. His best year was 1973, in which he led the National League in doubles, home runs, runs batted in, and slugging average. Stargell led the league in home runs one other season. He and Keith Hernandez were named co–Most Valuable Players in 1979. It was the first MVP tie ever, and Stargell, at 39, was the oldest ever to win the award. Stargell was selected to play in seven All-Star Games and was elected to the Baseball Hall of Fame in 1988.

The highest-ranking first baseman and the fourth-highest-ranking hitter overall was Dick Allen, with a PRG rating of 1.156. Allen played from 1963 to 1977: seven years with the Philadelphia Phillies, three years with the Chicago White Sox, two more years back with the Philadelphia Phillies, and one year with the Oakland Athletics. He was a very contentious person and became involved in a seemingly endless series of controversies, but he was also a very good hitter. In 1964 Allen was named Rookie of the Year. This was undoubtedly based on his hitting, for his fielding was definitely subpar. It wasn't his fault, however, because he was told to play third base even though he had never played that position before. The Phillies led the National League by six and a half games on September 20 but then lost 10 games in a row and ended up in a tie for second place.

Allen was blamed for the demise and was roundly criticized by both the Philadelphia fans and sportswriters.

Allen continued to hit well, but he also became involved in a series of controversial incidents—fist fighting, tardiness, drunkenness, and a self-inflicted injury to his throwing hand that required him to be moved to first base. He even became involved in a controversy over his name. Early in his career he was called "Richie." After several years, however, he suddenly announced he wanted to be called "Dick." "Richie," he maintained, was a little boy's name, and he demanded to be called the more masculine "Dick" instead.

Finally, Allen was traded by the Phillies, and then he was traded again and again. In his last nine years he played for five different teams. He had one great year (1972) with the White Sox, in which he led the American League in slugging average, on-base percentage, home runs, and runs batted in; finished third in batting average; and was named the Most Valuable Player. His production fell off in 1973, but in 1974 he again led the league in home runs and slugging average. Allen feuded with teammate Ron Santo and actually quit the team before the season ended. He returned to the Phillies for two more seasons; then, after being traded to Oakland and threatened with designated hitter status, he quit baseball for good. Allen was selected to play in seven All-Star Games but has never been elected to the Baseball Hall of Fame.

The second-highest-ranking first baseman and the fifth-highest-ranking hitter overall was Willie McCovey, with a PRG rating of 1.151. McCovey played from 1959 to 1980: 15 years with the San Francisco Giants, three years with the San Diego Padres, and four more years back with the Giants. He was one of the most popular San Francisco players ever. Willie was not his nickname but his real first name. His nickname was "Stretch" because, at 6 feet 4 inches tall, he had an unusually long stretch to reach balls thrown toward him at first base.

Willie McCovey, like Frank Robinson and Dick Allen, won the Rookie of the Year Award his first year in the National League. His opportunities to play first base were, however, limited during his first six years with the Giants. When entrenched first baseman Orlando Cepeda was injured and subsequently traded to the St. Louis Cardinals, McCovey became the regular first baseman. In his first season as a regular player for the Giants he led the league with 44 home runs.

McCovey peaked from 1968 to 1970 when he led the league in slugging average three times, in home runs and runs batted in twice, and in walks, doubles, and on-base percentage once each. He was voted the Most Valuable Player in 1969. McCovey was a feared hitter—he led the National League in intentional walks four times and finished his career with 260 intentional walks. In 1971 and 1972 he played only part-time because of injuries. McCovey was traded to San Diego in 1974 when he was 35 years old, and his batting statistics declined thereafter. He finished his career with 521 home runs, the second most ever for a left-handed batter in the National League. McCovey was selected to play in six All Star Games and was elected to the Baseball Hall of Fame in 1986.

The third-highest-ranking first baseman and the sixth-highest-ranking hitter overall was Harmon Killebrew, with a PRG rating of 1.148. Killebrew played from 1954 to 1975, spending 21 years with the Washington Senators / Minnesota Twins (seven years in Washington and 14 years in Minnesota) and one year with the Kansas City Athletics. Killebrew was called "Killer," but not because of his personality. He was respected as a quiet gentleman who didn't complain or argue with umpires. The nickname came from his very strong muscular body, which enabled him to hit so many home runs.

Killebrew was a quintessential one-dimensional slugger. He hit 573 home runs but had a lifetime batting average of only .256, struck out 1,699 times, hit into 243 double plays (the fifth most among the top 50 home run hitters of all time), and made 215 errors in the field. Killebrew hit more than 40 home runs eight times, leading the league in home runs six of those times. He ranks seventh all-time in home run percentage (7.03). Killebrew also led the league in walks four times, runs batted in three times, and on-base percentage and slugging average once each. He was selected to play in 11 All-Star Games and won one Most Valuable Player Award. Killebrew was elected to the Baseball Hall of Fame in 1984.

The second-highest-ranking left fielder and the seventh-highest-ranking hitter overall was Rico Carty, with a PRG rating of 1.116. Carty played from 1963 to 1979: eight years with the Milwaukee / Atlanta Braves, four years with the Cleveland Indians, and three years split with four other teams. He was born and learned to play baseball in the Dominican Republic. Rico was short for Ricardo, his first name. Carty hit more for average than for power. His best year was 1970, when he led the National League with a .366 batting average and a .454 on-base percentage. Carty also had a career-high 25 home runs and 100 runs batted in and was selected to play in the All-Star Game. Carty has never been elected to the Baseball Hall of Fame.

The fourth-highest-ranking first baseman and the eighth-highest-ranking player overall was Orlando Cepeda, with a PRG rating of 1.100. Cepeda played from 1958 to 1974: eight seasons with the San Francisco Giants, three seasons with the St. Louis Cardinals, four seasons with the Atlanta Braves, one season with the Boston Red Sox, and one season with the Kansas City / Oakland Athletics.

Cepeda was a right-handed power hitter. He was called "Baby Bull" for his power and also "Cha Cha" because he liked to listen to Latin music in the clubhouse. He hit well in his first seven years with San Francisco, with a combined batting average of .309 and an average of 32 home runs and 107 runs batted in per year. Cepeda injured his knee, however, and after surgery was traded to St. Louis in 1966. His best year was 1967, when he led the National League in runs batted in (111) and won the Most Valuable Player Award. Cepeda's statistics declined after that, but he finished his career with a lifetime batting average of .297, 1,365 runs batted in, and 379 home runs. Cepeda was elected to the Baseball Hall of Fame in 1999 by the Veterans Committee.

The third-highest-ranking left fielder and the ninth-highest-ranking player overall was Frank Howard, with a PRG rating of 1.088. Howard played from 1958 to 1973: seven years with the Los Angeles Dodgers, seven years with the Washington Senators, and one year each with the Texas Rangers and Detroit Tigers. Standing 6 feet 7 inches tall and weighing 255 pounds, he was one of the biggest and strongest players in the history of Major League Baseball. He was nicknamed "Hondo," after a John Wayne movie of the same name, and in Washington also became known as "The Capital Punisher."

Howard was strictly a power hitter. In his first full year with the Dodgers, his 23 home runs helped him win the Rookie of the Year award. Howard's productivity declined thereafter, however, and only after he was traded to Washington did it improve significantly. He averaged 34 home runs per year with a slugging average of .511 in his seven years with Washington. He led the league in home runs and slugging average in 1968. Howard's best year was 1970, when he led the league with 44 home runs, 126 runs batted in, and 132 walks. He was selected to play in four All-Star Games while with Washington. Frank Howard has never been elected to the Baseball Hall of Fame.

The fifth-highest-ranking first baseman and the 10th-ranked hitter overall was Norm Cash, with a PRG rating of 1.086. Cash played from 1958 to 1974, spending two seasons with the Chicago White Sox followed by 15 seasons with the Detroit Tigers. He was a left-handed power hitter who was also a rarity at first base—a good fielder with a strong arm.

Cash's best season by far was 1961, when he led the American League in hits, batting average, and on-base percentage. He also scored 119 runs, hit 41 home runs, and drove in 132 runs, but he still did not win the Most Valuable Player Award. This went to Roger Maris, who led the league in runs, runs batted in, and home runs (his 61 home runs broke Babe Ruth's single-season record of 60). Cash's statistics declined thereafter, and he never again led the American League in any major batting event. He did, however, enjoy a minor resurgence twice and was named Comeback Player of the Year both times. Cash was selected to play in four All-Star Games but has never been elected to the Baseball Hall of Fame.

The highest-ranking catcher and the 11th-ranked player overall was Johnny Bench, with a PRG rating of 1.084. Bench played from 1967 to 1983 and spent his entire 17-year major league career with the Cincinnati Reds. He lived a quiet life and was content to let his reputation ride on his performance on the field.

Bench, like many of the great players of this era, was named Rookie of the Year his first full year in the major leagues. Two years later he had the best year of his career, leading the National League with 45 home runs and 148 runs batted in, thus earning himself the Most Valuable Player Award. Two years after that he led the league with 40 home runs and 125 runs batted in and won his second Most Valuable Player Award. Two years after that he led the league with 129 runs batted in but finished second with only 33 home runs.

Johnny Bench was a very durable player. He set the National League record for games caught by a rookie catcher (154). Bench caught more than 140 games in nine seasons. He won 10 straight Gold Glove Awards and was selected to play in 14 All-Star Games, 13 of them consecutively. Bench's statistics did decline in the second half of his career, and he was transferred to first base and third base for his last three years. Bench was elected to the Baseball Hall of Fame in 1989.

The highest-ranking third baseman and the 17th-ranked hitter overall was Ron Santo, with a PRG rating of 1.073. Santo played from 1960 to 1974, spending 14 years with the Chicago Cubs and one final year with the Chicago White Sox. He was a very durable player, an excellent fielder, and hit for power. In one nine-year stretch he missed playing in only 16 games; won the Gold Glove Award five years in a row; led the league in walks four times, on-base percentage twice, and triples once; and averaged 27 home runs and 99 runs batted in. He finished his career with 342 home runs and 1,331 runs batted in. These were very good numbers for a third baseman. Ron Santo was selected to play in nine All-Star Games but has never been elected to the Baseball Hall of Fame.

The highest-ranking center fielder and the 25th-ranked hitter overall was Amos Otis, with a PRG rating of 1.048. Otis played from 1967 to 1984: two years with the New York Mets, 14 years with the Kansas City Royals, and one last year with the Pittsburgh Pirates. He was a balanced player, a good fielder, fairly fast, and had good but not outstanding hitting statistics. In one 10-year stretch he did average 29 doubles and 29 stolen bases per year, leading the league in doubles twice and stolen bases once (the only major batting events in which he ever led the league). Otis was selected to play in five All-Star Games but has never been elected to the Baseball Hall of Fame.

The highest-ranking second baseman and the 28th-ranked hitter overall was Joe Morgan, with a PRG rating of 1.011. Morgan played from 1963 to 1984: nine years with the Houston Astros, eight years with the Cincinnati Reds, a year back with Houston, two years with the San Francisco Giants, and a year each with the Philadelphia Phillies and Oakland Athletics. He was small, only 5 feet 7 inches tall and 160 pounds, but he more than compensated with his intelligence, aggressiveness, great fielding, and clutch hitting.

Morgan played well for Houston but blossomed into the best second baseman in baseball with Cincinnati. In his first six years with the Reds, he had a combined batting average of .302, on-base percentage of .430, and slugging average of .497 and averaged 22 home runs, 118 walks, and 60 stolen bases per year. Morgan led the league in on-base percentage four times and doubles, slugging average, walks, and runs once each. These are excellent statistics for a second baseman. He also won five straight Gold Glove Awards and two consecutive Most Valuable Player Awards. His statistics declined thereafter, but he continued to be selected in All-Star Games—a total of 10 for his entire major league career. Morgan was elected to the Baseball Hall of Fame in 1990.

The highest-ranking shortstop and the 35th-ranked hitter overall was Rico Petrocelli, with a PRG rating of .973. Petrocelli played from 1963 to 1976, spending his entire 13-year major league career with the Boston Red Sox. His nickname "Rico" was short for his first name Americo. He was a rarity for his time—a shortstop with some power. Petrocelli never led the league in a major batting event but did average 25 home runs and 80 runs batted in for one five-year stretch. He led the American League in fielding average three years, two of them in a row, but was moved to third base when his fielding declined. Petrocelli was selected to play in two All-Star Games but has never been elected to the Baseball Hall of Fame.

This was the era in which Hispanic and especially black players came into their own. In the previous era, Jackie Robinson broke the color barrier and Roy Campanella, Willie Mays, and Ernie Banks went on to prove that black players could compete with white players if given the chance. The black players of the Dead Ball Interval proved that they could outperform white players. Nearly half the players in table 6.2 were black or Hispanic. The top five were all black players: Hank Aaron, Frank Robinson, Willie Stargell, Dick Allen, and Willie McCovey. One other black player—Joe Morgan—was the best second baseman of this era. Eight highly rated Hispanic players—Rico Carty, Rod Carew, Cesar Cedeno, Orlando Cepeda, Roberto Clemente, Tony Oliva, Tony Perez, and Manny Sanguillen—were active in this era. As Hank Aaron's statement above indicates, most of the black players played in the National League. Only Frank Robinson played as much as half of his time in the American League. Most of the leading Hispanic players also played in the National League. Only Tony Oliva and Rod Carew spent their entire careers in the American League. Rico Carty split his career between the two leagues.

A total of 18 position players from this era were elected to the Baseball Hall of Fame. Eight of those players were the best hitters at their positions, or nearly so, and are listed in table 6.2. Five of the 10 players not listed in table 6.2—Rod Carew, Roberto Clemente, Tony Perez, Billy Williams, and Carl Yastrzemski—were known more for their hitting than for their fielding. Al Kaline was both a good hitter and a good fielder. Three of the 10 players not in table 6.2—Luis Aparicio, Bill Mazeroski, and Brooks Robinson—were known for their defensive play, and Robinson was the only one of the three to even once lead his league in a major batting event (118 runs batted in, in 1964). They were not particularly high average hitters (they all had lifetime batting averages in the .260s) or great sluggers (only one of them [Robinson] managed a slugging average as high as .401). Aparicio was also known for his speed on the base paths (506 career stolen bases). Lou Brock did not have an impressive hitting résumé for a left fielder, but he did lead his league in runs, doubles, and triples once each. This was largely attributable to his terrific speed. His 938 career stolen bases ranks second all-time to Rickey Henderson. Speed undoubtedly was the asset that earned Brock his election to the Hall of Fame.

7

The Live Ball Revived Era, 1977–1992

THE AGE OF MIKE SCHMIDT

Mike Schmidt was a great slugger, a record-setting fielder, a three-time National League Most Valuable Player, and the finest Philadelphia Phillie ever, according to a 1983 vote by Philadelphia fans. . . . He led the National League in home runs, a record eight times; only Ruth led his league more often. He slugged 30 or more home runs in a season 13 times, a figure surpassed only by Hank Aaron, and reached 35 homers 11 times, more often than anyone but Ruth. . . . He won 10 Gold Gloves, more than any third baseman except Brooks Robinson.

The Baseball Encyclopedia, p. 1005

The American League decided to try a designated hitter rule for a three-year period beginning in 1973. During the trial period, batting statistics improved and attendance increased. The league, not surprisingly, voted to make the designated hitter rule a permanent part of the rule book. In 1976 batting statistics declined across the board, but in 1977 they began to climb to levels not seen since the Live Ball Era (1921–1941). This climb was facilitated by a 1987 contraction in the upper level of the strike zone from the armpits to the midpoint between the top of the shoulders and the top of the pants. National League batting statistics also increased during this time, but, lacking a designated hitter rule, they lagged behind American League batting statistics. Beginning in 1993, batting statistics in both leagues began to climb even higher, inaugurating a fourth phase of live ball baseball. The era-by-era results are as follows (both leagues combined):

	AVG	OBP	SLG	OPS	HR%	ERPG
Dead Ball Interval (1963–1976)	.249	.329	.368	697	2.2	7.03
Live Ball Revived (1977–1992)	.259	.324	.387	711	2.4	7.68
Live Ball Enhanced (1993–2009)	.266	.335	.423	761	3.1	8.79

The top 10 hitters of the Live Ball Revived era for each of the average measurements of hitting performance are identified in table 7.1. The pattern of agreement and disagreement is similar to the one we have observed in the previous eras. If you look at the top 10 as a group, there is general agreement on five players who belong in the group. All of the measures agree on only one player (Will Clark), all except one measure agree on one other player (Schmidt), and all except two measures agree on three players (Tartabull, Strawberry, and Brett). Twenty-seven players share the top 10 rankings, and 16 players share the top five rankings. Both groups are the largest for all historical eras.

If you look at the specific ranking of the players within the top 10, there is also considerable disagreement. Mike Schmidt is ranked number 1 by five measures, but no other player is ranked the same by as many as three measures. This is the most disagreement by far that we have on specific player rankings for any historical era.

The extraordinary amount of disagreement between the various measures makes it even more imperative to follow the guidance in the Pregame Analysis in deciding on the leading hitters for this era. Table 7.2 summarizes the leading hitters of the Live Ball Revived era by the positions they played in the field. The highest-ranking third baseman and the highest-ranking hitter overall is Mike Schmidt, with a PRG rating of 1.165. Schmidt played from 1972 to 1989, spending his entire 18-year major league career with the Philadelphia Phillies. At the plate he was primarily a slugger: he batted over .300 only once and had a lifetime batting average of only .267. Schmidt was also a great-fielding third baseman—his 10 Gold Glove Awards are second only to the 16 awarded to Brooks Robinson.

Schmidt's home run exploits are noted in the above quotation: second to Babe Ruth for the eight seasons in which he led the National League in home runs, second to Hank Aaron for his 13 sea-

TABLE 7.1 The Top 10 Hitters in the LBR Era by Various Measures

Player	PRG	Advanced Weighted Measures			Basic Weighted Measures			Unweighted Measures	
		LSLR	RC/27	LWTS	TA	OPS	SLG	OBP	AVG
Mike Schmidt	1	1	2	2	1	1	1	6	—
Jose Canseco	2	2	—	4	5	4	3	—	—
Kevin Mitchell	3	—	—	—	7	3	2	—	—
Danny Tartabull	4	9	9	5	8	5	7	—	—
Will Clark	5	8	1	3	4	2	6	5	8
Darryl Strawberry	6	10	5	10	2	6	4	—	—
Kent Hrbek	7	—	—	—	—	—	—	—	—
Jim Rice	8	4	—	—	—	10	5	—	—
George Brett	9	7	10	5	—	8	9	—	6
Jack Clark	10	—	7	8	6	9	—	8	—
Eric Davis	(12)	—	8	9	3	—	—	—	—
Reggie Jackson	(14)	—	—	—	9	—	8	—	—
John Kruk	(17)	—	—	—	10	—	—	2	10
Kirby Puckett	(24)	5	—	—	—	—	—	—	3
Wade Boggs	(32)	6	3	1	—	7	—	1	2
Tony Gwynn	(35)	—	4	7	—	—	—	3	1
Paul Molitor	(37)	3	—	—	—	—	—	—	5
Other players[1]	—	—	—	—	—	—	—	—	—

[1] Ten other players ranked in the top 10 but in one measure only: Eddie Murray tied for #10 in LSLR; Pedro Guerrero #6 in RC/27; Fred Lynn #10 in SLG; Keith Hernandez tied for #3, Alvin Davis tied for #6, Kevin Seitzer #9, and Bobby Grich #10 in OBP; and Don Mattingly #4, Bill Madlock #7, and Al Oliver #9 in AVG.

sons in which he hit 30 or more home runs, and second to Babe Ruth for the 11 seasons in which he hit 35 or more home runs. He also led the National League in slugging average five times, in walks and runs batted in four times each, in on-base percentage three times, and in runs once. Schmidt had 100 or more walks in seven seasons and more than 100 RBI in nine seasons.

Schmidt was clearly the dominant hitter and the second-best-fielding third baseman of his era. He was selected to play in 12 All-Star Games and won three Most Valuable Player Awards. Schmidt

TABLE 7.2 Leading Hitters of the LBR Era by Position[1]

Position	Leading Hitter	Second-Leading Hitter and Others in the Top 10 Overall
Third base	Mike Schmidt — 1.165 (1)*	George Brett — 1.110 (9)*
Designated hitter	Jose Canseco — 1.161 (2)	Harold Baines — 1.084 (20)
Left field	Kevin Mitchell — 1.141 (3)	Jim Rice — 1.114 (8)*; Greg Luzinski — 1.099 (11)
Right field	Danny Tartabull — 1.139 (4)	Darryl Strawberry — 1.123 (6); Jack Clark — 1.108 (10)
First base	Will Clark — 1.130 (5)	Kent Hrbek — 1.118 (7)
Center field	Eric Davis — 1.096 (12)	Fred Lynn — 1.076 (22)
Catcher	Mickey Tettleton — 1.054 (31)	Ted Simmons — 1.049 (33)
Shortstop	Robin Yount — 1.024 (41)*	Cal Ripken Jr. — 1.009 (42)*
Second base	Bobby Grich — .989 (50)	Ryne Sandberg — .976 (55)*

* Indicates that a player was elected to the Baseball Hall of Fame.

[1] The number after each player's name is his potential runs per game (PRG) rating, and the number in parentheses is his PRG ranking for the LBR era.

was elected to the Baseball Hall of Fame in 1995, his first year of eligibility.

The highest-ranking designated hitter and the second-highest-ranking hitter overall was Jose Canseco, with a PRG rating of 1.161. Canseco played from 1985 to 2001: eight seasons with the Oakland Athletics, followed by two seasons each with the Texas Rangers and the Boston Red Sox, a season back with Oakland, a season with the Toronto Blue Jays, two seasons with the Tampa Bay Devil Rays, and one final season with the Chicago White Sox. Canseco's career straddles this era and the following era. He is included in this era because he had more plate appearances in this era. At 6 feet 4 inches tall and 240 pounds, he was one of the biggest players ever to play Major League Baseball. Canseco was also one of the most unfulfilled players of all time, largely because of his own antics, both on and off the field.

Canseco's early years with Oakland were very impressive. In 1986 he was named Rookie of the Year. In 1988 he became the first player in major league history to have at least 40 home runs and 40 stolen bases; led the league in home runs, runs batted in, and slugging average; and was voted the league's Most Valuable Player. In 1991 he again led the league in home runs. Unfortunately, Canseco also became involved in a series of off-the-field incidents, including speeding tickets, a deliberate automobile accident, and the illegal possession of firearms. Thus, when his statistics declined, he was traded to Texas. At Texas he had two on-the-field mishaps that further tarnished his image—letting a fly ball bounce off his head for a home run and missing most of a season because he tore a ligament while foolishly attempting to pitch in a game.

Canseco was very good at three things: stealing bases, hitting home runs, and driving in runs. He finished his career with 200 stolen bases, 462 home runs, and 1,407 runs batted in. Canseco ranks 18th all-time in home run percentage (6.55) and 25th all-time in runs batted in percentage (19.9). His 200 stolen bases are 11th among the 50 all-time home run leaders, and he would have had many more had he not been bothered by injuries in the second half of his career. Canseco was selected to play in six All-Star Games. After he retired, he admitted to having used steroids when he played. Canseco has never been elected to the Baseball Hall of Fame.

The highest-ranking left fielder and the third-highest-ranking player overall was Kevin Mitchell, with a PRG rating of 1.141. Mitchell played from 1984 to 1999 and was one of the most-traveled players of his era. He spent 13 seasons with eight different teams, four in each league: two seasons with the New York Mets, part of a season with the San Diego Padres, four-plus seasons with the San Francisco Giants, one season with the Seattle Mariners, two seasons with the Cincinnati Reds, a season split between the Boston Red Sox and the Cincinnati Reds, a season with the Cleveland Indians, and finally one last season with the Oakland Athletics.

Mitchell was a good hitter and was versatile in the field. In the 1986 season he was a great help to the Mets with his timely hitting and by playing six different positions in the field. Unfortunately, Mitchell never lived up to his early promise because he was constantly playing with various injuries. In only 6 of his 13 seasons did he play more than 100 games, and during his last four seasons he averaged only 34 games each. Mitchell's best season was 1989 with San Francisco, when he led the National League in home runs, runs batted in, slugging average, and intentional walks and won the National League's Most Valuable Player Award. Mitchell was selected to play in two All-Star Games but has never been elected to the Baseball Hall of Fame.

The highest-ranking right fielder and the fourth-highest-ranking hitter overall was Danny Tartabull, with a PRG rating of 1.139. Tartabull played from 1984 to 1997: three seasons with the Seattle Mariners, five seasons with the Kansas City Royals, three-plus seasons with the New York Yankees, and two-plus seasons split with the Oakland Athletics, Chicago White Sox, and Philadelphia Phillies. Danny was short for Danielo, his first name.

Tartabull was primarily a slugger. He had 25 or more home runs and 100 or more runs batted in five times and had 25 or more home runs two other seasons. His best years were those in Kansas City. Tartabull's best year was 1991, when he had a .316 batting average, 31 home runs, 100 runs batted in, and played in his only All-Star Game. It was the only year in which he led his league in a major batting event—slugging average (.593). Tartabull was not a particularly good outfielder or base runner and has never been elected to the Baseball Hall of Fame.

The highest-ranking first baseman and the fifth-highest-ranking hitter overall was Will Clark, with a PRG rating of 1.130. Clark played from 1986 to 2000: eight years with the San Francisco Giants, five years with the Texas Rangers, and two final years split between the Baltimore Orioles and the St. Louis Cardinals. Will was short for William, his first name. He had two nicknames: "Will the Thrill" for his aggressive style of play and "The Natural" for his smooth swing.

Clark was both a percentage hitter and a slugger, as well as a good-fielding first baseman. Sluggers, especially sluggers who play first base, are not usually known for their fielding ability. In 1991 Clark accomplished the rare feat of leading the league in slugging average and winning the Gold Glove Award. He batted .300 or over

11 times, had 23 or more home runs five times, and had more than 100 runs batted in four times. He led the league in runs, runs batted in, walks, and slugging average once each. Will Clark was selected to play in six All-Star Games but has never been elected to the Baseball Hall of Fame.

The second-highest-ranking right fielder and the sixth-highest-ranking hitter overall was Darryl Strawberry, with a PRG rating of 1.123. Strawberry played from 1983 to 1999: eight seasons with the New York Mets, three with the Los Angeles Dodgers, one with the San Francisco Giants, and five with the New York Yankees. His nickname was "Straw," short for Strawberry.

Strawberry was primarily a slugger—he hit over .300 only once and finished with a lifetime batting average of only .259. His best seasons were with the Mets. In his first season he won the National League Rookie of the Year Award. Strawberry had 24 or more home runs 10 times (eight of them with the Mets) and had three seasons (all of them with the Mets) with 37 or more home runs and more than 100 runs batted in. His very best season was 1988, in which he led the league in home runs and slugging average. Early in Strawberry's career he had some speed and stole 25 or more bases five years in a row. He was selected to play in eight All-Star Games (seven of them when he was with the Mets).

Strawberry's batting statistics may have been better if not for several problems—colon cancer, drug abuse, legal issues, and injuries—that limited his playing time. In his last eight seasons he played in 100 or more games only once. Strawberry has never been elected to the Baseball Hall of Fame.

The second-highest-ranking first baseman and the seventh-ranked hitter overall was Kent Hrbek, with a PRG rating of 1.118. Hrbek played from 1981 to 1994, spending his entire 14-year major league career with the Minnesota Twins. He never led the league in a major batting event but did have some good seasons. Hrbek hit more for power than for average. He hit over .300 only three times and finished with a lifetime batting average of .282. Hrbek hit 25 or more home runs six times and finished his career with a total of 293. He drove in 90 or more runs five times and finished his career with 1,086 runs batted in. His best year was 1984, in which he hit .311 with 27 home runs and 107 runs batted in. Kent Hrbek was selected to play in only one All-Star Game and has not been elected to the Baseball Hall of Fame.

The second-highest-ranking left fielder and the eighth-highest-ranking hitter overall was Jim Rice, with a PRG rating of 1.114. Rice played from 1974 to 1989 and spent his entire 16-year major league career with the Boston Red Sox. In the International League, Rice had won the Triple Crown and was named both Rookie of the Year and Most Valuable Player. His arrival in Boston was eagerly anticipated by the Red Sox, and he did not disappoint them. Rice had a great rookie season with the Red Sox—a .309 batting average, 22 home runs, and 102 runs batted in—but his teammate, Fred Lynn, was named Rookie of the Year and Most Valuable Player, with a .331 batting average, 21 home runs, and 105 runs batted in.

Jim Rice was primarily a slugger, but he also hit well for average. He batted over .300 seven times and finished his career with a .298 batting average. Rice had more than 100 runs batted in eight times, and six of those times he also had 27 or more home runs. He led the league in home runs three times, runs batted in and slugging average twice, and doubles and hits once. Rice's best year was 1978, when he led the American League in hits, triples, home runs, runs batted in, and slugging average and won the Most Valuable Player Award. He was not a fast runner and ranks third all-time in grounding into double plays (315). Rice was selected to play in eight All-Star Games and was elected to the Baseball Hall of Fame in 2009.

The second-highest-ranking third baseman and the ninth-highest-ranking hitter overall was George Brett, with a PRG rating of 1.110. Brett played from 1973 to 1993, spending his entire 21-year career with the Kansas City Royals. He was a left-handed batter who hit for average (.305 lifetime) and also had some power (317 career home runs). Brett led the American League in hits, triples, batting average, and slugging average three times each, doubles twice, and on-base percentage once. He had 3,154 career hits and 1,595 runs batted in. His best year was 1980, when he led the league with a .390 batting average, .445 on-base percentage, and .664 slugging average and received the Most Valuable Player Award. Brett's .390 batting average remains the second highest in the major leagues—Tony Gwynn hit .394 in 1994—since Ted Williams batted .406 in 1941. He is the only player in baseball history to win batting titles in three decades—1976, 1980, and 1990. Brett was selected to play in 12 All-Star Games, 11 of them consecutively. He was elected to the Baseball Hall of Fame in 1999.

The third-highest-ranking right fielder and the 10th-highest-ranking hitter overall was Jack Clark, with a PRG rating of 1.108. Clark had an 18-year major league career with several teams: ten years with the San Francisco Giants, three years with the St. Louis Cardinals, a year with the New York Yankees, two years with the San Diego Padres, and finally two years with the Boston Red Sox. Jack was his first name, not a nickname.

Early in his career Clark was a good all-around player—a power hitter who ran the bases and fielded his position well. After a while, however, he was slowed by various injuries and became strictly a slugger. Clark led his league in walks three times and on-base percentage and slugging average once each. He had 25 or more home runs eight times, 20 or more doubles seven times, and 90 or more runs batted in five times. Clark ended his career with 340 home runs, 332 doubles, and 1,180 runs batted in. He was selected to play in four All-Star Games but has never been elected to the Baseball Hall of Fame.

The highest-ranking center fielder and the 12th-ranked hitter overall was Eric Davis, with a PRG rating of 1.096. Davis played from 1984 to 2001, spending eight years with the Cincinnati Reds followed by nine years with five other teams (two years with the Los Angeles Dodgers, one year with the Detroit Tigers, one year back with Cincinnati, two years with the Baltimore Orioles, two years with the St. Louis Cardinals, and finally one year with the San Francisco Giants), thus making him one of the most-traveled players of this era. He was called "Eric the Red" after the old Viking explorer of that name.

Eric Davis was a slugger with speed. He had seven seasons with 24 or more home runs, and in six of those seasons he had 21 or more stolen bases. This was amazing because he suffered from various injuries throughout his career and even retired for one season before returning to the game for his last six seasons. Davis was also a very good fielder and won three consecutive Gold Glove Awards. He never led his league in any major batting event but when healthy was among the leaders in home runs and was always a threat to steal a base. Davis was selected to play in only two All-Star Games and has not been elected to the Baseball Hall of Fame.

The highest-ranking catcher and the 31st-ranked hitter overall was Mickey Tettleton, with a PRG rating of 1.054. Tettleton played from 1984 to 1997, spending his first four seasons with the Oakland Athletics, followed by three seasons with the Baltimore Orioles, four seasons with the Detroit Tigers, and finally three seasons with the Texas Rangers. He was primarily a catcher with the first three teams but was switched to designated hitter by Texas.

Tettleton was a switch-hitter with home run power. From 1989 to 1996 (including the strike-shortened 1994 season), he averaged 26 home runs per year, including four seasons with 31 or more. He had four seasons with 83 or more runs batted in, including one with 110. His ranking as the number 1 catcher of his era was attributable to his power rather than his averages—a lifetime batting average of only .241 and an on-base percentage of only .369. Ted Simmons had a much higher batting average, but not nearly as much home run power.

The only major batting category in which Tettleton ever led the American League was walks—122 in 1992. He was selected to play in two All-Star Games but has never been elected to the Baseball Hall of Fame.

The highest-ranking shortstop and the 41st-ranked hitter overall was Robin Yount, with a PRG rating of 1.024. Yount played from 1974 to 1993, spending his entire 20-year career with the Milwaukee Brewers. For his first 11 seasons he played shortstop, and for the last nine seasons he played center field. He is one of only three major league players ever to win a Most Valuable Player Award at two different positions. Hank Greenberg and Stan Musial did it at the less demanding positions of first base and left field.

Yount was a good all-around steady shortstop—a good hitter,

fielder, and base runner—who consistently led his teammates by example. In 1982 he had one of the best seasons ever for a shortstop. He led the American League in hits (210), doubles (42), total bases (367), slugging average (.578), and on-base plus slugging (957) and just missed the batting title with a .331 average versus .332 for Willie Wilson. Yount led the American League in doubles one other season and in triples two seasons. He averaged 29 doubles a season for his entire career. Yount was selected to play in three All-Star Games and is a member of the exclusive 3,000-hit club. Robin Yount was elected to the Baseball Hall of Fame in 1999.

The second-highest-ranking shortstop and the 42nd-ranked hitter overall was Cal Ripken Jr., with a PRG rating of 1.009. Ripken played from 1981 to 2001, spending his entire 21-year major league career with the Baltimore Orioles. "Cal," of course, was short for Calvin, both his and his father's first name.

Cal Ripken Jr. holds one of the most difficult-to-break records in the history of baseball—playing in 2,632 straight games. Ripken's streak started in 1982, the year he was named American League Rookie of the Year. The following year he won the Most Valuable Player Award, leading the American League in runs, hits, and doubles, but never again led the league in a major batting event. Yet during the 17-year streak, Ripken averaged 32 doubles, 23 home runs, and 89 runs batted in per year. Ripken was also a good fielder, leading all shortstops in fielding five times and winning two consecutive Gold Glove Awards. He won two American League Most Valuable Player Awards and was selected to play in 18 straight All-Star Games. Cal Ripken was elected to the Baseball Hall of Fame in 2002, his first year of eligibility.

The highest-ranking second baseman and the 50th-ranked hitter overall was Bobby Grich, with a PRG rating of only .989. He played from 1970 to 1986, spending his entire 17-year major league career with only two teams: seven years with the Baltimore Orioles followed by ten years with the California Angels. "Bobby" was short for Robert, his first name.

Grich was known less for his hitting than for his fielding. When he was with Baltimore, he won four consecutive Gold Glove Awards and frequently led the league in various defensive events such as put-outs, assists, and double plays. Grich and shortstop Mark Belanger were the best double-play combination in baseball. He also hit well for a second baseman when he was with Baltimore. Grich led the American League in home runs and slugging average in the strike-shortened 1981 season. In his five years as Baltimore's regular second baseman, he averaged 89 walks, 83 runs, and 27 doubles per season.

After the 1976 season, Grich took advantage of free agency and signed a lucrative contract with the Angels. The following year, however, he hurt his back and was able to play in only 52 games. In Grich's next eight years as the Angels regular second baseman, his batting statistics declined—he averaged 69 walks, 67 runs, and 20 doubles per season. His best year was 1979, with 30 doubles, 30 home runs, and 101 runs batted in. Grich's injured back also affected his fielding statistics—no more Mark Belanger, no more Gold Glove Awards, and no more leading the American League in defensive events. Grich was selected to play in six All-Star Games but has not been elected to the Baseball Hall of Fame.

A total of 18 position players from this era were elected to the Baseball Hall of Fame. Nine of those players—George Brett, Wade Boggs, Tony Gwynn, Reggie Jackson, Paul Molitor, Eddie Murray, Jim Rice, Mike Schmidt, and Dave Winfield—were probably chosen based mainly on their hitting accomplishments, although several of them were also accomplished fielders. Nine other players—Gary Carter, Andre Dawson, Carlton Fisk, Rickey Henderson, Cal Ripken Jr., Ryne Sandberg, Kirby Puckett, Ozzie Smith, and Robin Yount—were probably chosen based on a variety of factors. Ripken probably owes his election to the Hall of Fame mostly for his fantastic streak of 2,632 consecutive games and his "nice guy" popular image, although he was also a good-fielding and good-hitting shortstop. Despite an assortment of injuries, Fisk held the all-time record for the number of games caught (2,226) until 2009, when it was broken by Ivan Rodriguez. Yount was a steady all-around player at two positions. Sandberg was known more for his fielding than for his hitting—he won nine consecutive Gold Glove Awards at second base and had several errorless streaks in the course of his career. Carter was one of the best defensive catchers of his time. Many believe Ozzie Smith to be the best-fielding shortstop ever. Speed was an important consideration in the election of Rickey Henderson, who holds the all-time record for stolen bases (1406). Kirby Puckett and Andre Dawson were good hitters and fielders, so both factors probably figured in their election.

8

The Live Ball Enhanced Era, 1993–2009

THE AGE OF UNCERTAINTY

Everyone involved in baseball over the past two decades—Commission-
ers, club officials, the Players Association, and players—shares to some
extent in the responsibility for the steroids era. There was a collective
failure to recognize the problem as it emerged and to deal with it early
on. As a result, an environment developed in which illegal use became
widespread.

The Mitchell Commission Report, December 13, 2007

Before pursuing the issue of steroids in baseball, it is necessary to set
the statistical framework for the era in which the players played and
the issue emerged. In 1993 and again in 1994 there were across-the-
board increases in batting statistics in both leagues to levels significantly higher than those achieved in the preceding era. These levels
were pretty much retained for the remainder of the era, although
there were some seasonal ups and downs. The American League statistics continued to be a little higher than the National League statistics because of the designated hitter rule in the American League. In
2001, however, the National League had a higher home run percentage than the American League. That was the year Barry Bonds set
the single-season record of 73 home runs and Sammy Sosa became
the first player with 60 or more home runs in three seasons. A comparison of live ball eras follows (both leagues combined):

	AVG	OBP	SLG	OPS	HR%	ERPG
Live Ball Era (1921–1941)	.281	.344	.398	742	1.4	8.26
Live Ball Continued (1942–1962)	.259	.331	.382	712	2.1	7.66
Live Ball Revived (1977–1992)	.259	.324	.387	711	2.4	7.68
Live Ball Enhanced (1993–2009)	.266	.335	.423	761	3.1	8.79

The one consistent trend in these data is the era-by-era increase in home run percentage. Since the home run percentage increases consistently while the earned runs per game declines and remains approximately the same for two eras and then increases, home runs cannot be the determining factor in the production of runs. An increase in home run percentage does not necessarily lead to a comparable increase in the number of earned runs scored.

Now we can consider the steroids issue. In the 1990s, a series of reports indicated that baseball players were using drugs to enhance their performance. The reports appeared credible because the physiognomy of players was changing and the rate of home run hitting was increasing. Finally, in 2003, Major League Baseball attempted to determine the extent of drug abuse. An anonymous test was conducted—the names of the players testing positive would remain confidential, but the overall results would determine whether some action was required. Because more than 5% of the players tested positive, a regular unannounced testing program was scheduled to begin in the 2004 season.

In September of 2003, the Bay Area Laboratory Co-Operative (BALCO) facility in suburban San Francisco was raided by federal agents investigating the use of performance-enhancing drugs in sports. Greg Anderson, an employee of BALCO, was also the personal trainer of Barry Bonds. In December 2003, Bonds testified before a federal grand jury following up on information obtained in the BALCO raid. He testified that he used some substances given to him by Anderson but that he did not know they were steroids. In February 2004, Anderson and three other employees of BALCO pled guilty to charges of participating in the distribution of performance-enhancing drugs to athletes. Anderson was sentenced to three months in prison to be followed by three more months of in-house confinement. The other BALCO employees were sentenced to varying terms of imprisonment, house confinement, and probation.

In March 2006, Bud Selig, the commissioner of Major League Baseball, asked George J. Mitchell, a former U.S. Senate majority party leader, to lead an investigation of the alleged use of performance-enhancing drugs in baseball. In July 2006, Greg Anderson was found in contempt of court for refusing to testify about Bonds before the federal grand jury and was sent back to prison. In April 2007, Kirk Radomski, a former New York Mets clubhouse attendant, pled guilty to distributing performance-enhancing drugs to many athletes and money laundering between 1995 and 2005. His sentence was postponed because he was cooperating with the Mitchell Commission investigation. In October 2007, track athlete Marion Jones pled guilty to charges of lying to federal prosecutors and admitted that she used steroids. Jones was the sixth person convicted in the BALCO investigation (a fifth BALCO employee not referred to above, chemist Patrick Arnold, had also been convicted); all six had pled guilty and accepted plea bargains before their cases went to trial. Jones's track coach, Trevor Graham, was also convicted of lying to federal investigators and sentenced to one year of house arrest.

In November 2007, Barry Bonds was indicted on charges of perjury and obstruction of justice in connection with his testimony before the federal grand jury in 2003. Greg Anderson was immediately released from prison. The government had apparently obtained the agreement of another witness or witnesses to testify, and the testimony of Anderson was no longer required. The trial of Bonds will probably not start until early 2011—more than three years since he was indicted and eight years since his alleged felonious testimony before the grand jury.

The Mitchell Commission released its report on December 13, 2007. Over the preceding 19 months, the commission had conducted a thorough investigation. It reviewed over 100,000 pages of documents and over 20,000 electronic records from the baseball commissioner's office and major league teams. It also interviewed more than

600 people connected, in some official capacity, with Major League Baseball teams. The commission wanted to interview more than 50 players, but, almost without exception, the players declined.

The Mitchell Commission report named 91 players as users of steroids and other performance-enhancing drugs but recommended that they not be disciplined for their conduct except when it was "so serious that discipline is necessary to maintain the integrity of the game." Preferring to look forward instead of backward, the commission recommended that Major League Baseball (1) establish an enhanced year-round testing program to be administered by an independent company, (2) investigate allegations of drug use in the absence of positive tests, and (3) educate players about the risks involved in using performance-enhancing drugs.

In February 2008, Radomski was sentenced to five years probation for distributing steroids to athletes and money laundering. He received a light sentence because he had cooperated with the Mitchell Commission investigation. Later that month, Roger Clemens testified before the House of Representatives Committee on Oversight and Government Reform. His former trainer, Brian McNamee, had accused Clemens of using steroids and human growth hormone. McNamee repeated his accusations and Clemens repeated his denials before the committee. Most of the Democratic members appeared to believe McNamee, and most of the Republican members appeared to believe Clemens. An affidavit from Clemens's former teammate Andy Pettitte appeared to favor McNamee. Henry Waxman, the chairman of the committee, said that "it's impossible to believe this is a simple misunderstanding. Someone isn't telling the truth." The committee then asked the Department of Justice to investigate whether Clemens committed perjury in his testimony before the committee. In January 2009 a federal grand jury heard evidence pertaining to allegations that Clemens perjured himself in testimony before Congress. In August 2010, Clemens was indicted on three counts of making false statements and on two counts of perjury in his testimony before Congress.

The Mitchell Commission recommendations have been acted on, although not to the complete satisfaction of all concerned. In

January 2008, Major League Baseball established a department in the commissioner's office to investigate drug use in baseball. In April 2008, Major League Baseball players and owners agreed on more frequent drug testing and an increase in the authority of the outside administrator in charge of the testing program. All players implicated in the Mitchell report were given amnesty. The commissioner indicated that baseball management people in the report would be required to perform some act of public service in lieu of punishment. The Players Association would participate in Major League Baseball's program to educate youth about performance-enhancing drugs and contribute the sum of $200,000 for that purpose.

It will take some time before we can assess whether these actions have resulted in a significant decrease in the use of performance-enhancing drugs in baseball. We will have to wait until the Clemens and Bonds trials are completed and the new league drug testing program proceeds. In February 2009, it was reported that Alex Rodriguez and Barry Bonds had tested positive for steroids back in 2003. This revelation was leaked to the press in the wake of other press reports that Rodriguez had used steroids early in his career. Rodriguez then confirmed that he had. In May of 2009, it was announced that Manny Ramirez tested positive for a banned substance and was fined $7.65 million and suspended for 50 games. In June, it was reported that Sammy Sosa had tested positive in 2003. In July, it was revealed that David Ortiz and Manny Ramirez had also tested positive in 2003! Thus, the steroids fallout continues and the list of prominent players who have used steroids continues to grow. What revelations await us in the months and years ahead?

Since two-thirds of the players named in the Mitchell report were everyday players, rather than pitchers, the impact of the use of steroids should have been reflected more in batting statistics than in pitching statistics. It is interesting to look at the batting statistics, but before we do, let's make some assumptions. The commission was presented with an unofficial code of silence on the part of the players. Almost without exception, the players declined to talk with the commission, and it had no subpoena power to compel the players to testify. Players involved with steroids naturally did not want to admit their guilt, nor did they want to tell the commission what they knew about their teammates. If the players had been more forthcoming, it is reasonable to assume that more players would have been implicated in the use of steroids. The commission was able to see only the part of the steroid iceberg above water. One could infer, however, from the tone of the report and the remedies it recommended, that the commission suspected that the number of players using steroids was much greater than those it was able to name in the report. If so, there should have been a significant impact on batting statistics.

The Mitchell report says that the steroids era emerged about two decades ago—that would make it around the 1987 season, give or take a few years either way. Since steroids are supposed to build up your strength and to maintain your endurance, they should enable the user to hit more home runs than the nonuser. If the steroid era began to emerge two decades ago, and the use of steroids was as widespread as the report says, we should see an impact on home run statistics.

During the 1985, 1986, and 1987 seasons, the percentage of home runs (per at bat) increased dramatically. The 1987 rate of 3.094 was the highest ever up to that time, and the 1987 season became known as the "Year of the Home Run." Of the 61 everyday players in the Mitchell report, only four were active in 1985, eight in 1986, and 11 in 1987, and five of the 1987 players had fewer than 250 at bats. Thus, the surge in home runs could not be blamed solely on the Mitchell report players. Some attributed the surge in home runs to an unusually hot summer, but major league officials evidently thought it was something more lasting. They expanded the strike zone for the 1988 season, and the percentage of home runs declined and remained at pre-1985 levels for several seasons.

During the 1993 and 1994 seasons, the percentage of home runs once again increased dramatically—the 1994 rate of 2.998 was close to the 1987 rate, the second highest ever up to that time. This time no adjustment was made by league officials, and the percentage of home runs remained relatively stable until the 1999 and 2000 seasons, when a third surge brought the two highest home run rates ever—3.307 in 1999 and 3.403 in 2000. After the 2000 season, the percentage of home runs gradually declined to 2.962 in 2008—the lowest season since 1994—and increased slightly to 3.040 in 2009.

Table 8.1 summarizes the percentage of home runs for everyday Mitchell report players and all other players by five-year periods from 1983 to 2007. If the steroids era emerged in 1988, the percentage of home runs hit in the first five years of the steroids era actually declined by 13% from the percentage of home runs hit in the last five years of the pre-steroids era. This was true for Mitchell report players and for all other players—the decline was actually greater for the Mitchell report players (23%) than for all other players (15%). During the second five-year period of the steroids era, the percentage of home runs increased significantly. The increase was greater for all other players (29%) than for the Mitchell report players (25%). During the third five-year period of the steroids era, the percentage of home runs increased a little more, and the increase, once again, was greater for all of the other players (10%) than for the Mitchell report players (3%). During the fourth and last five-year period of the steroids era, the percentage of home runs declined, but that decline was much greater for Mitchell report players (18%) than for all of the other players (less than 1%).

Thus, whenever the percentage of home runs increased, the increase was greater for other players than for Mitchell report players, and whenever the percentage of home runs decreased, the decrease was less for the other players than for the Mitchell report players. This suggests that steroids may not have that great an effect or, if they do, that many of the non–Mitchell report players were also on steroids. In each five-year period, the percentage of home runs is greater for the Mitchell report players than for the other players. This suggests that steroids do matter. Putting both suggestions together, we can conclude that steroids probably helped both the Mitchell report players and many of the other players. It has been speculated by some that as many as 50% of the major league players of this era may have used steroids at one time or another.

Another supposed "benefit" from steroids is the restorative effect, or endurance. It is interesting to note, however, that only two Mitchell report players—Barry Bonds and Rafael Palmeiro—are among

TABLE 8.1 Mitchell Report Players' and Other Players' Home Runs by Five-Year Periods

Years	Mitchell Report Players			Other Players			All Players			Ratio of MRPs to OPs
	AB	HR	%	AB	HR	%	AB	HR	%	
1983–1987	6,539	308	4.710	711,104	18,124	2.549	717,643	18,432	2.568	1.85
Change		−23			−15			−13		
1988–1992	34,441	1,254	3.641	679,579	14,747	2.170	714,020	16,001	2.241	1.68
Change		+25			+29			+31		
1993–1997	56,787	2,584	4.550	659,284	18,435	2.796	716,071	21,019	2.935	1.63
Change		+3			+10			+10		
1998–2002	71,555	3,354	4.687	761,804	23,448	3.078	833,359	26,802	3.216	1.52
Change		−18			NC			−3		
2003–2007	40,482	1,555	3.841	795,066	24,463	3.077	835,548	26,018	3.114	1.25

the nine all-time great home run hitters (500 or more career home runs) who have played into their 40s. Bonds hit 59 home runs for a home run percentage of 7.88 after the year he turned 40. Ted Williams was second with 39 home runs for a home run percentage of 6.70 after the year he turned 40. None of the other seven—Aaron, Reggie Jackson, Mays, McCovey, Palmeiro, Frank Robinson, and Murray—had as many as 25 home runs or a home run percentage of 5.0 or more after the year they turned 40. Thus, steroids may help rejuvenate the body on a short-term day-to-day basis, but this evidence suggests that steroids may not help to extend a player's career. It suggests instead that steroids may actually work to limit the length of a player's career.

The top 10 hitters (both Mitchell report players and other players) of the Live Ball Enhanced Era for each of the average measures of batting performance are identified in table 8.2. The pattern of agreement and disagreement between measures is similar to that we have seen before. If you look at the top 10 as a group, there is gen-

TABLE 8.2 The Top 10 Hitters in the LBEE by Various Measures

Player	PRG	Advanced Weighted Measures			Basic Weighted Measures			Unweighted Measures	
		LSLR	RC/27	LWTS	TA	OPS	SLG	OBP	AVG
Albert Pujols	1	1	2	2	2	1	1	3	1
Manny Ramirez	2	4	6	4	5	3	3	7	6
Barry Bonds	3	2	1	1	1	2	2	1	—
Lance Berkman	4	5	9	7	6	7	—	6	—
Frank Thomas	5	6	7	6	5	6	—	4	—
Mark McGwire	6	—	4	9	4	5	4	—	—
Todd Helton	7	7	5	3	3	4	7	2	3
Alex Rodriguez	8	3	—	5	9	9	5	—	—
Albert Belle	9	10	—	—	—	—	10	—	—
Jim Thome	10	—	—	—	7	10	9	—	—
Larry Walker	(12)	—	3	—	8	8	8	—	8
Jeff Bagwell	(16)	8	8	8	10	—	—	8	—
Vladimir Guerrero	(17)	9	—	—	—	—	6	—	4
Chipper Jones	(20)	—	—	10	—	—	—	9	—
Edgar Martinez	(21)	—	10	—	—	—	—	5	9
Other players[1]	—	—	—	—	—	—	—	—	—

[1] Five other players are ranked in the top 10 but in one measure only: Jason Giambi #10 in OBP; and Ichiro Suzuki #2, Derek Jeter #5, Nomar Garciaparra #7, and Mike Piazza #10 in AVG.

eral agreement on seven players who belong there. All of the measures agree on three players (Pujols, Ramirez, and Helton), all except one measure agree on one player (Bonds), and all except two measures agree on two other players (Berkman and Thomas). Twenty players share the top 10 rankings, and 13 players share the top five rankings.

If you look at the specific rankings of the players within the top 10, there is a lot of disagreement. Four measures agree that Bonds should be ranked number 1, but three other measures agree that he should be ranked number 2. The situation is similar for Albert Pujols. Five measures agree that he should be ranked number 1, but three other measures agree that he should be ranked number 2. Two other players are ranked the same by four measures: Walker and

Bagwell. Three other players—Thomas, McGwire, and Helton—are ranked the same by three measures.

What we have, one more time, is a dilemma. If you want to know who the leading hitters of an era are, you have to choose among the various models or combination of models. To solve this dilemma, just recall what was said in favor of the PRG measure in the Pregame Analysis and subsequent chapters.

Identification of the leading hitters of this era also needs to be approached in the light of the steroids issue. Seven of the 16 leading hitters (44%) have been implicated: Ramirez, Bonds, McGwire, Rodriguez, Ortiz, Gonzalez, and Giambi. If these seven players were excluded, the PRG rankings would be #1 Pujols, #2 Berkman, #3 Thomas, #4 Helton, #5 Belle, #6 Thome, #7 Walker, #8 Delgado, #9 Bagwell, and #10 Guerrero. It will be very interesting—and perhaps controversial, divisive, and painful—to see the impact of the steroids issue on future elections to the Baseball Hall of Fame. Mark McGwire has been passed over three times, and Barry Bonds will become eligible in two more years. Since four of the implicated players are still active, this process will take some time to play itself out.

Table 8.3 summarizes the leading hitters of the Live Ball Enhanced Era by the positions they played in the field. Five of the 10 highest-ranking hitters are first basemen. The highest-ranking first baseman and the highest-ranking hitter overall is Albert Pujols, with a PRG rating of 1.333. He has spent his entire nine-year career with the St. Louis Cardinals. Pujols is both a percentage hitter and a slugger. He has led the National League in runs scored four times, in slugging average three times, in on-base percentage twice, and in hits, doubles, batting average, and home runs once each. Pujols has received a Rookie of the Year Award and three Most Valuable Player Awards and has been selected to play in eight All-Star Games, including the last seven in a row. Since Pujols is only 30 years old, his great career batting statistics will probably improve in the future.

The highest-ranking left fielder and the second-highest-ranking hitter overall is Manny Ramirez, with a PRG rating of 1.321. "Manny" is short for Manuel, his first name. Ramirez has played eight years with the Cleveland Indians, seven and two-thirds years

TABLE 8.3 Leading Hitters of the LBEE by Position[1]

Position	Leading Hitter	Second-Leading Hitter and Others in the Top 10 Overall
First base	Albert Pujols — 1.333 (1)	Lance Berkman — 1.269 (4); Mark McGwire — 1.265 (6); Todd Helton — 1.251 (7); Jim Thome — 1.236 (10)
Left field	Manny Ramirez — 1.321 (2)	Barry Bonds — 1.291 (3); Albert Belle — 1.245 (9)
Designated hitter	Frank Thomas — 1.268 (5)	David Ortiz — 1.234 (11)
Shortstop	Alex Rodriguez — 1.248 (8)	Nomar Garciaparra — 1.141 (36)
Right field	Larry Walker — 1.230 (12)	Vladimir Guerrero — 1.219 (17)
Third base	Chipper Jones — 1.203 (20)	Scott Rolen — 1.140 (37)
Catcher	Mike Piazza — 1.188 (22)	Jorge Posada — 1.149 (31)
Center field	Ken Griffey Jr. — 1.173 (23)	Jim Edmonds — 1.158 (28)
Second base	Jeff Kent — 1.128 (45)	Alfonso Soriano — 1.028 (60)

[1] The number after each player's name is his potential runs per game (PRG) rating, and the number in parentheses is his PRG ranking for the LBEE.

with the Boston Red Sox, and one and one-third years with the Los Angeles Dodgers. Ramirez has led the American League in slugging average and on-base percentage three times each and in batting average, runs batted in, and home runs once each. A big factor in his high PRG rating is a consistently high number of runs batted in. Ramirez is a veritable RBI machine—more than 100 runs batted in for 12 of his 16 seasons as a regular player, including nine seasons in a row. Ramirez's career RBI percentage (22.7) is the highest of all active players. Ramirez has been selected to play in 12 All-Star Games, including 10 in a row. Since he is 38 years old, his batting statistics will probably decline in the years ahead. His accomplishments are suspect, however, because he has been tainted by the steroids controversy. He was fined $7.65 million and suspended for 50 games in May of 2009 because he used a banned substance. After he returned to the lineup, it was revealed that he had also tested positively for the use of a performance-enhancing substance in the 2003 major league test.

The second-highest-ranking left fielder and the third-highest-ranking hitter overall is Barry Bonds, with a PRG rating of 1.291. Bonds spent seven years with the Pittsburgh Pirates, followed by 15 years with the San Francisco Giants. Barry Bonds has some of the most amazing statistics of any player ever to play the game. In his 22-year career, he led the National League in walks 12 times, on-base percentage 10 times, slugging average and on-base plus slugging seven times each, home runs and batting average twice each, and runs and runs batted in once each. He was selected to play in 14 All-Star Games and won eight Gold Glove Awards and seven Most Valuable Player Awards. Bonds ranks first all-time in career home runs (762) and is third in home run percentage (7.7). Perhaps most amazingly, his statistics actually increased as he got older. The prime years for most players are in their 20s. For Bonds it was not just in his 30s, but in his late 30s! Consider the following statistics:

Age	AVG	OBP	SLG	OPS	HR%
22–28	.275	.380	.503	883	5.38
29–35	.302	.439	.617	1056	7.82
36–40	.339	.535	.781	1316	12.16
41–43	.274	.464	.561	1025	7.88

The increase in his statistics in his early 30s was remarkable enough, but the increase in his statistics in his late 30s amounts to a quantum leap, which is followed by a return in his early 40s to the level of his early-30s statistics.

It has not yet been proven that Bonds used steroids, but many baseball fans believe that he did. If Bonds is convicted in his perjury and obstruction of justice trial, the vast majority of baseball fans will be convinced that he used steroids. How else, they will say, did Bonds achieve such amazing numbers for a player in his late 30s? A guilty verdict in his trial would not be conclusive evidence that he used steroids, but that plus the circumstantial evidence that he did would be so great that virtually all impartial observers would be convinced. It's a shame, because his statistics may have been good enough to rank him with the all-time greats without the quantum leap of his late 30s. Bonds, along with all other players in the Mitchell report, has been given an amnesty from being disciplined by the commissioner's office. However, Bonds faces other penalties. His image has been tarnished, even if he is exonerated by the courts. Readers will have a mental asterisk by his name as they peruse the record books. He could be found guilty by the courts and spend time in jail. And his future election to the Baseball Hall of Fame has already been jeopardized. Look at what has happened to Mark McGwire on this subject.

The second-highest-ranking first baseman and the fourth-highest-ranking hitter overall is Lance Berkman, with a PRG rating of 1.269. Berkman has spent his entire 11-year major league career with the Houston Astros. He is a switch-hitter who hits more for power than for average. Berkman has a lifetime slugging average of .555 versus a lifetime batting average of .299. He has averaged 32 home runs per year for the last nine years. The only major hitting events in which Berkman has led the league are doubles (twice) and runs batted in (once). He has had more than 100 runs batted in six times, more than 100 runs five times, and 30 or more home runs five times. Berkman has been selected to play in five All-Star Games. Since he is only 34 years old, his batting statistics may improve somewhat in the future.

The highest-ranking designated hitter and the fifth-highest-ranking hitter overall was Frank Thomas, with a PRG rating of 1.268. Thomas spent 16 seasons with the Chicago White Sox, followed by one season with the Oakland Athletics, one season with the Toronto Blue Jays, and one season split between Toronto and Oakland. He was one of the biggest players in the history of the game (6 feet 5 inches and 275 pounds) but a surprisingly balanced hitter with a lifetime batting average of .301 and 521 career home runs. Thomas led the American League in walks and on-base percentage four times each and in doubles, runs, batting average, and slugging average once each. In a 10-season stretch (1991–2000), he had a combined batting average of .320 and averaged 34 home runs and 115 runs batted in per season. He received two consecutive Most Valuable Player Awards and was selected to play in five All-Star Games. Thomas retired at the end of the 2008 season.

The third-highest-ranking first baseman and the sixth-highest-ranking hitter overall was Mark McGwire, with a PRG rating of 1.265. McGwire played from 1986 to 2001, spending 11-plus seasons with the Oakland Athletics followed by four-plus seasons with the St. Louis Cardinals. At 6 feet 5 inches and 225 pounds, he was sometimes referred to as "Big Mac." McGwire was a quintessential slugger: in his rookie year at Oakland he hit 49 home runs, had a slugging average of .618, and won the Rookie of the Year Award. He teamed with another Oakland slugger, Jose Canseco, forming a duo known as the "Bash Brothers."

McGwire's slugging statistics are impressive. He holds the all-time record for home run percentage (HR ÷ AB) at 9.4%. In one spectacular four-year span (1996–99), he had a home run percentage of 13.7 and a slugging average of .706. McGwire led his league in home runs and slugging average four times, in walks and on-base percentage twice, and in runs batted in once. He was selected to play in 12 All-Star Games—nine for American League and three for National League teams. In 1998, McGwire and Sammy Sosa raced each other in pursuit of Roger Maris's home run record (61)—Sosa finished with 66, but McGwire overtook him with 70. Three years later Barry Bonds overtook McGwire with 73.

McGwire, like Bonds, has been tainted by the steroids controversy. He admitted taking a powdered supplement called androstenedione. The substance was banned in some sports leagues, but not in baseball. McGwire said he took the supplement to "prevent" injuries and stopped using it after the 1998 season.

The prime years for most players are in their 20s, but McGwire, like Bonds, had his prime years in his 30s. McGwire's statistics for batting average, on-base percentage, slugging average, on-base plus slugging, and home run percentage were all significantly higher in his 30s than in his 20s. Whether or not the use of androstenedione was responsible for his late-career surge is problematic. The fact that he admitted to using androstenedione, however, is probably the reason that McGwire has not been admitted to the Baseball Hall of Fame.

The fourth-highest-ranking first baseman and the seventh-highest-ranking hitter overall is Todd Helton, with a PRG rating of

1.251. Helton has spent his entire 13-year major league career with the Colorado Rockies. He is a percentage hitter who also hits for power. His hitting has been helped by the thin air in Denver, but his numbers are, nevertheless, quite impressive. In one seven-year stretch (1998–2004) Helton had a combined batting average of .340 and 25 or more home runs each year, and in five of those years he also had more than 100 runs batted in. His best season was in 2000, when he led the National League in hits, doubles, runs batted in, batting average, on-base percentage, and slugging average. The only other year in which Helton led the league in a major hitting event was in 2005, when he led the league in on-base percentage. Helton has been selected to play in five All-Star Games. Since Helton is nearly 37 years old, his career batting statistics may decline as his career winds down.

The highest-ranking shortstop and the eighth-highest-ranking hitter overall is Alex Rodriguez, with a PRG rating of 1.248. Rodriguez has played seven seasons with the Seattle Mariners, three seasons with the Texas Rangers, and the last six seasons with the New York Yankees. Rodriguez was quite big for a shortstop (6 feet 3 inches and 225 pounds), which undoubtedly was a factor when the Yankees moved him from shortstop to third base. He is listed as a shortstop here because he has played many more games at shortstop than at third base. His nickname is "ARod"—"A" for the first letter of his first name and "Rod" for the first three letters of his last name.

Rodriguez is a balanced hitter, with a lifetime batting average of .305 and 583 career home runs. He has led the American League in runs and home runs five times each, in slugging average four times, in runs batted in twice, and in hits, doubles, and batting average once each. In Rodriguez's 14 seasons as a regular he has a combined batting average of .307 and has averaged 41 home runs and 120 runs batted in. This is quite remarkable. He has received three Most Valuable Player Awards and has been selected to play in 11 All-Star Games. Rodriguez is only 35 years old and probably has several good seasons left to enhance his already impressive batting statistics.

The third-highest-ranking left fielder and the ninth-highest-ranking hitter overall was Albert Belle, with a PRG rating of 1.245. Belle played from 1989 to 2000: eight years with the Cleveland Indians, followed by two years each with the Chicago White Sox and the Baltimore Orioles. Belle was a right-handed power hitter (381 home runs and a slugging average of .564), but he also hit respectably for average (lifetime batting average of .295). Belle led the American League in runs batted in three times, slugging average twice, and runs, doubles, and home runs once each. His best year was 1995, when he led the league in runs, doubles, home runs, runs batted in, and slugging average. He lost the Most Valuable Player Award to Mo Vaughn, who tied Belle in runs batted in but trailed Belle significantly in the other aforementioned events. The fact that Vaughn was popular and Belle was controversial may have had something to do with the voting. Belle's career was marked with repeated confrontations with fans, umpires, management, and the media. Belle was selected to play in five consecutive All-Star Games but has never been elected to the Baseball Hall of Fame.

The fifth-highest-ranking first baseman and the 10th-ranked hitter overall is Jim Thome, with a PRG rating of 1.236. He has spent 12 years with the Cleveland Indians, followed by three years with the Philadelphia Phillies and four years with the Chicago White Sox. Thome is a big (6 feet 4 inches and 245 pounds) power-hitting left-handed batter. Thome has led his league in walks three times and in home runs and slugging average once each. He was a feared hitter. Among sluggers he ranks sixth all-time in walk percentage (one walk for every 5.84 plate appearances). In one nine-year stretch he averaged 114 walks, 145 hits, 41 home runs, and 111 runs batted in. Thome's best year was 2002, when he had 52 home runs, 118 runs batted in, and a .304 batting average, .445 on-base percentage, and .677 slugging average. Thome has been selected to play in five All-Star Games.

The second-highest-ranking designated hitter and the 11th-ranked hitter overall is David Ortiz, with a PRG of 1.234. Ortiz has spent his entire major league career with two teams—six seasons with the Minnesota Twins and seven seasons with the Boston Red Sox. At 6 feet 4 inches and 230 pounds, "Big Papi" endeared himself with Red Sox fans because of his role in banishing the "curse of the Bambino" by helping them win two World Series championships.

Ortiz played over 100 games only once (130 games in 2000) in his six seasons in Minnesota. In his seven seasons with Boston, he has led the league in runs batted in and walks twice and in home runs and on-base percentage once each. Ortiz is primarily a slugger, but he also hits for average. While with Boston he has a slugging average of .578 and a batting average of .288. In the 2004 and 2007 World Series years, he hit several dramatic game-winning home runs, giving him a reputation for being a clutch hitter. Ortiz has averaged 119 runs batted in, 87 walks, and 37 home runs per season while with Boston. In 2008 he was injured and missed part of the season, and early in 2009 he had his worst slump since joining the Red Sox. In July 2009, it was revealed that he had tested positively for performance-enhancing drugs in baseball's 2003 test. He has been selected to play in five All-Star Games.

The highest-ranking third baseman and the 20th-ranked hitter overall is Chipper Jones, with a PRG rating of 1.203. Jones has played his entire 16-year major league career with the Atlanta Braves. "Chipper," of course, is a nickname—his real first name is Larry. When Larry was growing up, his father was the high school baseball coach and people thought that his son was a chip off the old block.

Jones is a switch-hitter with balanced batting statistics—a lifetime batting average of .307 and 426 career home runs. He led the National League in batting average and on-base percentage in 2008, the only time he has ever led the league in a major batting event. Jones has batted over .300 10 times, has hit 25 or more home runs 10 times, and has driven in 100 or more runs nine times. These are not spectacular numbers, but they are very good numbers, especially for a third baseman. He has received one Most Valuable Player Award and has been selected to play in six All-Star Games. Since Jones is 38 years old and his statistics declined in 2009, they may decline further in the years ahead.

The highest-ranking catcher and the 22nd-ranked hitter overall is Mike Piazza, with a PRG rating of 1.188. Piazza spent six seasons with the Los Angeles Dodgers, split a season between the Dodgers, Florida Marlins, and New York Mets, and then spent seven full seasons with the Mets, one season with the San Diego Padres, and one season with the Oakland Athletics. Piazza was a right-handed batter who hit both for average and for power, with a lifetime batting average of .308 and 427 career home runs. He never led his league in a major batting event, nor does he rank high in any all-time career batting events. However, in his first 10 years as a regular he had a combined batting average of .322 and averaged 35 home runs and 107 runs batted in per year. Piazza did this while catching an average of 130 games per year, a very impressive accomplishment. He won the National League Rookie of the Year Award in 1993 and was selected to play in 12 All-Star Games, including 10 of them in a row. He retired at the end of the 2007 season.

The highest-ranking center fielder and the 23rd-ranked hitter overall was Ken Griffey Jr., with a PRG rating of 1.173. Griffey played his first 11 seasons with the Seattle Mariners, followed by eight seasons with the Cincinnati Reds, one season split between Cincinnati and the Chicago White Sox, and one season back with Seattle. He was much bigger than the average center fielder (6 feet 3 inches and 205 pounds) but played the position extremely well. Griffey was a left-handed batter who was more of a slugger than average hitter—630 career home runs but a lifetime batting average of only .285. He led his league in home runs four times and in runs, runs batted in, and slugging average once each. In Seattle his combined batting average was .299, compared with .270 in Cincinnati. Griffey's power numbers, however, declined only slightly. His home run percentage in Seattle was 6.8 compared with 6.3 in Cincinnati, and his RBI percentage was 19.8 in Seattle compared with 18.0 in Cincinnati. Griffey received one Most Valuable Player Award and was selected to play in 13 All-Star Games, including 11 in a row. He retired early in the 2010 season.

The highest-ranking second baseman and the 45th-ranked hitter overall was Jeff Kent, with a PRG rating of 1.128. "Jeff" is short for his first name Jeffrey. Kent was a well-traveled player. He split his first year between the Toronto Blue Jays and the New York Mets, spent three years with the Mets, split a year between the Mets and the Cleveland Indians, and then spent six years with the San Francisco Giants, two years with the Houston Astros, and four years with the Los Angeles Dodgers.

Kent was a big (6 feet 2 inches and 210 pounds) second baseman whose hitting compensated for any shortcomings in the field. He never led his league in a major batting event, nor does he rank high in any all-time career batting events. But he was a steady player. Kent had a lifetime batting average of .290 while amassing 24 or more doubles 14 times, 20 or more home runs 12 times, and 80 or more runs batted in 10 times. He won one Most Valuable Player Award and was selected to play in five All-Star Games. Kent retired at the end of the 2008 season.

The statistical accomplishments of these players are impressive, but they were compiled at a time when the integrity of the game has been questioned. The Mitchell Commission found steroids to be the cause of a serious problem for baseball. Fourteen of the leading hitters discussed in this book were linked to steroids by the Mitchell Commission or other sources—the seven mentioned on page 128 plus Jose Canseco, Lenny Dykstra, Troy Glaus, Rafael Palmeiro, Gary Sheffield, Sammy Sosa, and Miguel Tejada. These players were enabled, in effect, to enhance an already live ball in an effort to gain riches and baseball immortality. Baseball statistics, the lifeblood of the game, have been contaminated and are no longer reliable indicators of the performance of some players. Unfortunately, we don't know which players. We don't know for sure who used steroids and for how long, and we don't know who did not use steroids. We are uncertain about the recent past and about the immediate future. The great legacy of this era is uncertainty.

PART II

The Ultimate Lineup Card

* 9 *

Fielding a Team of Great Hitters

A president's cabinet is like a baseball team. The members of a baseball team are chosen for their qualifications to play a particular defensive position, but they are also expected to take their turn at bat and help the team score runs. The members of a president's cabinet are chosen for their expertise in the cabinet department which they represent so they can defend the department from outside criticism. They are also expected to take part in the discussion of general issues and help formulate administration policy. Thus, members of a president's cabinet have to play both defense and offense, just like the members of a baseball team.

Frederick E. Taylor, May 1996

....

In chapters 1–8 we rated and ranked the leading hitters from a historical perspective—era by era. In this chapter we will combine the results of those chapters and rate and rank the all-time leading hitters from a position-by-position perspective, that is, for all 134 years of Major League Baseball history from 1876 to 2009. Before proceeding, however, we need to consider some realities about hitting potential position by position.

Size, Speed, Power, and Position
Different positions in the field place different physical demands on players that affect their hitting potential. Infielders need to be quicker than outfielders because they are closer to the batters and must react more quickly to a variety of critical situations. Since quickness is

related to size—the smaller the player, the quicker he is—infielders tend to be quicker and smaller than outfielders. Since power is also related to size—the bigger the player, the more power he has—infielders tend not to be power hitters. Fielding is more important in relation to hitting for shortstops and second basemen than for first basemen and third basemen. The latter two have fewer balls hit to them because they play closer to the baselines. Third basemen, however, have to be quicker than first basemen because they have less time to react to sharply hit balls. They are equidistant from the plate, but there are more right-handed batters than left-handed batters. On less sharply hit balls third basemen can range farther from the bag than first basemen, who have to be ever mindful of their primary responsibility to take throws at first base. First basemen also have to make fewer and shorter throws than third basemen. Thus, fielding is more important in relation to hitting for third basemen than for first basemen. First basemen, therefore, tend to be the biggest and best hitters and third basemen the second-biggest and second-best hitters in the infield.

Since outfielders do not have to be as quick as infielders, they tend to be bigger than infielders, and because they are bigger, they tend to be better hitters than infielders. Fielding is more important in relation to hitting for center fielders than for left and right fielders because they have more balls hit to them and more ground to cover. Thus, left fielders and right fielders tend to be bigger and better hitters than center fielders.

There is an important exception to the generalization that outfielders as a group tend to be bigger and better hitters than infielders as a group. This is true with regard to left fielders and right fielders versus second basemen, third basemen, and shortstops, but not with regard to center fielders versus first basemen. In this case the generalization should be reversed. Fielding is more important versus hitting for center fielders than for first basemen because center fielders have much more ground to cover than the relatively stationary first basemen and have more critical throws to make. Center fielders, therefore, tend to be smaller and better fielders and first basemen tend to be bigger and better hitters.

The requirements for a catcher are the most varied of all. Their hitting has to be weighed not only against their fielding (feet, arms, and gloves) but also against their handling of pitchers. Only rarely, therefore, are catchers among the leading hitters. Designated hitters have no fielding responsibilities and tend to be among the biggest players and best hitters on American League teams.

There are, of course, some individual exceptions to these generalizations about size, speed, power, and position. We are not talking about an ironclad rule, but about tendencies and averages. There is a reason, therefore, why many first basemen and few shortstops and second basemen are among the all-time great sluggers. If you happen to find a shortstop or second baseman among the all-time great sluggers, celebrate his performance because it was achieved in spite of the conflicting physical demands of his position.

Some information has been gathered that supports these generalizations. Table 9.1 summarizes the size and speed of 458 highly rated players ranked by position for different periods of time. There appears to be an inverse relationship between size and speed for each period of time. First basemen and catchers are the biggest and slowest players, while center fielders, shortstops, and second basemen are the smallest and fastest players. Third basemen, left fielders, and right fielders are ranked in the middle in both size and speed. Designated hitters have been excluded from the table because the designated hitter rule has been in effect in only one league and for only little more than a quarter of a century.

The top 50 hitters of all time, by position and according to various measures, are listed in table 9.2. More outfielders than infielders are included among the top 50 hitters for all measures except one. The differences between the number of outfielders and infielders are considerable—at least 1.4 outfielders for every infielder and as many as 1.8 outfielders for every infielder for two of the measures. Among the outfielders, center fielders lag behind left fielders and right fielders. Center fielders have the fewest players in five of the nine models and do not have the most players in any model. Among the infielders, first basemen are far ahead of the other positions. There are more first basemen in the top 50 than all the other infield positions com-

TABLE 9.1 The Size and Speed of Players Ranked by Position over Time[1]

	1876–1941		1942–2006		2007	
Position	Size	Speed	Size	Speed	Size	Speed
First basemen	1	7	1	7	1	7
Catchers	2	8	2	8	2	8
Right fielders	3	3	5	4	3	6
Third basemen	4	5	4	6	4	5
Left fielders	5	6	3	5	5	4
Center fielders	6	1	6	1	6	1
Shortstops	7	4	7	2	7	2
Second basemen	7	2	8	3	7	3

[1] Based on a review of the size (average weight) and speed (stolen bases per game) of 458 highly rated players.

bined for all measures except one. Overall, there are more first basemen among the top 50 hitters than at any other position in eight of the nine models. Catchers lag far behind all other positions. No model has more than one catcher, and two models have none at all. There are comparatively few designated hitters among the top 50 because the designated hitter rule has been in effect only in the American League for a little more than a quarter of a century.

One word of caution is needed before we proceed. We are about to compile lists of the top 10 hitting players at each position, not lists of the top 10 all-around players at each position. To compile a list of the top 10 all-around players at any position would require an acceptable fielding statistic and a way of balancing it with hitting statistics. When the National League was established, it adopted the fielding average (PO+A) ÷ (PO+A+E) as its official fielding statistic. It has since had many detractors, but no one has come up with an agreeable alternative. In a 1954 *Life* magazine article, Branch Rickey said, "there is nothing on earth anybody can do with fielding." Some progress has been made since Rickey's day, but there is still no general agreement on which statistics should be used to evaluate fielding. Even more difficult is the task of integrating fielding statistics with hitting statistics to come up with the all-around worth of baseball players. The most publicized attempt at this is Bill James's Win Shares system, but it is complex and has yet to gain a significant fol-

TABLE 9.2 The Top 50 Hitters of All Time by Position According to Various Measures

Position	PRG	Advanced Weighted Measures			Basic Weighted Measures			Unweighted Measures	
		LSLR	RC/27	LWTS	TA	OPS	SLG	OBP	AVG
First basemen	14	12	12	12	13	13	13	11	9
Left fielders	12	11	13	9	10	11	10	10	11
Right fielders	11	9	8	12	9	11	11	7	12
SUBTOTAL	37	32	33	33	32	35	34	28	32
Center fielders	5	9	10	10	9	8	8	6	9
Third basemen	2	3	2	4	5	1	2	5	3
SUBTOTAL	7	12	12	14	14	9	10	11	12
Designated hitters	3	1	2	2	3	3	2	2	—
Second basemen	1	1	1	1	1	1	1	6	4
Shortstops	1	3	1	1	1	1	2	2	1
Catchers	1	1	1	—	—	1	1	1	1
SUBTOTAL	6	6	5	4	5	6	6	11	6
ALL PLAYERS[1]	50	50	50	51	51	50	50	50	50

[1] Some totals may exceed 50 because of ties for the 50th position.

lowing in the baseball community. Thus, this book compiles lists of the leading hitters at each position rather than the leading all-around players at each position.

One of the problems with rating hitters who played at different times is to account for differences in the overall level of play and the level of competition during their respective careers. The performance of all players is partly a reflection of the tides prevailing at the time they played. High tides raise all boats and low tides lower all boats. What do you do if two hitters have the same rating but one played at a time when league ratings were above average and the other played at a time when league ratings were below average? It doesn't seem fair to give each hitter the same rating because the hitter who played when the league ratings were lower has probably accomplished more. Some external factor was probably tilting the balance between pitchers and batters in favor of pitchers. Perhaps, for example, the baseball itself was "dead," the height of the mound was higher, or the strike zone was bigger. Under any of these circumstances the hitter who played when league ratings were lower has accomplished more than the hitter who played when league ratings were higher.

The process of adjusting for differences in the balance between pitchers and batters at different times is called "normalization." One way to normalize a player's rating is to divide it by the league average during the years in which he played. Another way is to adjust a player's average by the amount of the difference between the league average during his career and the all-time average. If the league average was higher than the all-time average, you subtract the difference from the player's average; if the league average was lower than the all-time average, you add the difference to the player's average. Whichever of the two ways is chosen to normalize the player's rating, the ranking of the players is the same.

Some may object to this approach by taking the position that hitters are better today because they are reaping the benefits of progress in the science of hitting. Look, they say, at the relentless advance of the home run era by era since the Dead Ball Era. Actually, the home run rate (HR/AB) has increased steadily era by era, but PRG rates and other measures of hitting have not done so (see part I). Furthermore, while there has been progress in the science of hitting, there has also been progress in the art of pitching. Progress may proceed by fits and starts, but it cannot be confined to one of these areas to the exclusion of the other for very long. Hitting and pitching, offense and defense, are constantly adjusting and readjusting to one another—in every game, every season, and every era.

One final reminder is worthwhile before proceeding with the position-by-position discussion. Each of the position-by-position sections will repeat the table summarizing the top 10 hitters according to the various ways of measuring them. The objective of this format is twofold: first, to show the great differences in the modular rankings of players; and second, to confront the reader with the necessity of making a choice between the models. If you want to identify the best hitters for an era or for a position, you have to choose a measurement or a combination of measurements. That choice should be reasonable (i.e., supported by objective criteria) and consistent (i.e., the same for all eras and positions). The measurement used for evaluating first basemen, for example, should also be used for left fielders and right fielders. The process of scoring runs applies to all batters regardless of what position they play in the field. Therefore, whatever measurement you conclude is best should be applied to all batters.

As you look at these tables, it is important to remember what was said in the Pregame Analysis—the various measures are not of equal value. What we have in these tables is a hierarchy of measurement groups—three advanced weighted measures in group one, three basic weighted measures in group two, and two unweighted measures in group three. This hierarchy is based on a combination of logic and statistical testing. As you go from left to right, the measures decrease in reliability. The advanced weighted measures are the most reliable, the basic weighted measures are less reliable, and the unweighted measures are the least reliable of all. Similarly, within each of these measurement groups reliability decreases as you go from left to right. Statistical testing indicates that all of the advanced weighted measures are good predictors of runs per game—LSLR is the best, followed closely by RC/27 and LWTS, in that order.

And how does the PRG measure relate to this hierarchy of measurement groups? The answer depends on whether you are talking about teams or players. At the team level, the PRG measure tested the best of all. That does not mean, however, that the PRG player measure is also, ipso facto, the best of all. It may be the best or it may not be the best. It is impossible to say for certain whether it is or not.

We cannot say whether it is or not because the PRG team and player measures differ somewhat—unlike the measures in the hierarchy. It was possible to test the PRG team measures against team runs because both could be paired at the team level. It is not possible to test the player PRG measures against runs because individual players cannot be paired with runs. Most runs are scored by the cooperation of two or more players. The player PRG measures do, however, contain a high degree of reliability because about 62.5% of the bases accumulated by the average player are calculated with exactly the same factors used in the team PRG measure, and the remaining 37.5% of the bases were calculated with factors extrapolated from the team PRG measure. The changes, moreover, were small, more like a minor surgical procedure than a major heart operation.

It all boils down to the answer to another question. Was the margin of PRG primacy at the team level greater or lesser than any loss in PRG accuracy at the player level? The margin of PRG primacy over the average measures and basic weighted measures was almost certainly sufficient to offset any loss, but this is less certain with regard to the advanced weighted measures. It is also important to keep in mind that there may be no loss at all, and maybe even a gain, in PRG accuracy at the player level. To repeat what has been said many times before, there is no way to prove that one of these player measures is better than the other. But that does not mean that one cannot argue for one of them based on theory, reason, logic, etc. The reader should remember that the argument in favor of the PRG player measure applies to each of the positions addressed below, even though the explicit argument will not be repeated ad nauseam.

Shortstops

Fielding is very important in relation to hitting for shortstops because they have to react quickly in a variety of critical situations. Because quickness is related to size—the smaller the player, the quicker he is—shortstops tend to be small. Because size, in turn, is related to power—the smaller the player, the less powerful he is—shortstops tend not to be big hitters. Shortstops, therefore, are generally quicker, smaller, and less productive hitters than most other position players.

Table 9.1 indicates that shortstops (along with second basemen) are, on average, the smallest position players on the field in both the past and the present. From 1876 to 1941 the leading shortstops averaged 5 feet 10 inches and 170 pounds, from 1942 to 2006 they averaged 6 feet and 180 pounds, and in 2007 they averaged 6 feet and 192 pounds. Shortstops have become bigger, but they remain, along with second basemen, the smallest position players on the field because the players in all of the other positions have also become bigger. Table 9.1 also indicates that shortstops are the second fastest players on the field. In each of the three periods of time, the leading shortstops stole more bases per game than players at any other position except center field. Table 9.2 indicates that shortstops are not big hitters. Only one or two shortstops are among the top 50 hitters of all time in any of the various measures for ranking hitters except LSLR, which includes three shortstops.

The top 10 hitting shortstops of all time according to the various measures of hitting performance are identified in table 9.3. Alex Rodriguez played shortstop for Seattle and Texas but moved to third

TABLE 9.3 The Top 10 Hitting Shortstops of All Time According to Various Measures

Player	PRG	Advanced Weighted Measures			Basic Weighted Measures			Unweighted Measures	
		LSLR	RC/27	LWTS	TA	OPS	SLG	OBP	AVG
Alex Rodriguez	1	1	1	1	1	1	1	7	10
Joe Cronin	2	—	7	8	5	4	5	8	—
Nomar Garciaparra	3	3	8	5	6	2	2	—	5
Honus Wagner	4	6	2	2	2	5	6	4	1
Hugh Jennings	5	8	3	7	4	—	—	6	7
Arky Vaughan	6	7	4	3	3	3	9	1	2
Ed McKean	7	4	5	10	10	—	—	—	—
Ernie Banks	8	—	—	—	—	7	3	—	—
Miguel Tejada	9	10	—	—	—	10	4	—	—
Glenn Wright	10	—	—	—	—	—	10	—	—
George Davis	(11)	—	6	—	9	—	—	—	—
Joe Sewell	(12)	—	—	—	—	—	—	5	6
Derek Jeter	(13)	2	10	6	7	6	8	9	3
Luke Appling	(15)	—	—	—	—	—	—	2	8
Sam Wise	(21)	—	9	9	—	—	—	—	—
Barry Larkin	(22)	—	—	3	8	9	—	—	—
Vern Stephens	(26)	—	—	—	—	8	7	—	—
Johnny Pesky	(34)	—	—	—	—	—	—	3	9
Other players[1]	—	—	—	—	—	—	—	—	—

[1] Five other shortstops were ranked in the top 10 but in one measure only: Jimmy Rollins #5 and Rafael Furcal #9 in LSLR, Cal Ripken tied for #10 in SLG, Lou Boudreau #10 in OBP, and Cecil Travis #4 in AVG.

base for the New York Yankees. He is listed here as a shortstop because, through the 2009 season, he had played more games at shortstop than at third base. Because he is only 35 years old, he will probably end his career having played fewer games at shortstop than at third base. For now, however, he is rated as a shortstop.

The pattern of agreement and disagreement between the measures in table 9.3 is similar to what we saw in the historical eras. If you look at the top 10 shortstops as a group, there is general agreement on seven players who belong there. All of the measures agree on three players (Rodriguez, Wagner, and Vaughan), all except one measure agree on two players (Garciaparra and Jeter), and all except two measures agree on two other players (Cronin and Jennings). Twenty-three players share the top 10 rankings, and 16 players share the top five rankings.

If you look at the specific rankings of the shortstops within the top 10, there is even more disagreement. Seven measures agree that Rodriguez should be ranked number 1, three measures agree that Wagner should be ranked number 2, and three measures agree that Vaughan should be ranked number 3. None of the remaining 20 shortstops have more than two measures that agree on their specific ranking.

Table 9.4 summarizes the normalized PRG ratings for the top 10 hitting shortstops of all time. Players from the first two historical eras—the Era of Constant Change and the Live Ball Interval—have not been included in this table and in similar tables in the player position sections that follow, because the conditions under which they played were so much different from those in subsequent eras. As was mentioned in chapter 1, the rules affecting the interaction between pitchers and batters changed very frequently and players made many more errors during the Era of Constant Change. Batting statistics were greatly distorted, making it impossible for fair comparisons between players in this era and players in subsequent eras. The pitcher-batter rules stabilized during the Live Ball Interval, but significant differences remained between that era and those that followed. Home plate, for example, remained a 12-inch square and was not changed to its present size and shape (a five-sided figure 17

TABLE 9.4 Normalizing the PRG Ratings of the Top 10 Hitting Shortstops

Before Normalization			Normalization		After Normalization		
			LG				
Rank	Rating	Player	PRG	NORM[1]	Rank	Rating	Player
1	1.248	Alex Rodriguez	.986	1.266	1	1.407	Honus Wagner*
2	1.169	Joe Cronin	.960	1.218	2	1.266	Alex Rodriguez
3	1.141	Nomar Garciaparra	.971	1.175	3	1.229	Ernie Banks*
4	1.140	Honus Wagner	.810	1.407	4	1.218	Joe Cronin*
5	1.086	Arky Vaughan	.924	1.175	5	1.175	Arky Vaughan*
6	1.074	Ernie Banks	.874	1.229	6	1.175	Nomar Garciaparra
7	1.068	Miguel Tejada	.978	1.092	7	1.133	Glenn Wright
8	1.058	Glenn Wright	.934	1.133	8	1.118	Robin Yount*[2]
9	1.041	Joe Sewell	.955	1.090	9	1.117	Lou Boudreau*[3]
10	1.038	Derek Jeter	.985	1.054	10	1.098	Barry Larkin[4]

* Indicates that a player was elected to the Baseball Hall of Fame.
[1] Player's PRG rating divided by his league's PRG rating for the years in which he played.
[2] Before normalization Yount was ranked 12th with a PRG rating of 1.024 vs. a league average of .916.
[3] Before normalization Boudreau was ranked 16th with a PRG rating of 1.011 vs. a league average of .905.
[4] Before normalization Larkin was ranked 18th with a PRG rating of .998 vs. a league average of .909.

inches wide) until 1900. Players made fewer errors in the Live Ball Interval than before, but the error rate remained much higher than it became later on. The resultant distortion of batting statistics was less in the Live Ball Interval but remained significant enough to make it impossible to make fair comparisons between the players of that era and those of the preceding era and especially the subsequent eras.

We identified the great players of the Era of Constant Change and the Live Ball Interval in chapters 1 and 2 and included four leading shortstops from those eras—Hugh Jennings, Ed McKean, George Davis, and Sam Wise—in table 9.3, a compilation based on raw data unadjusted for differences in the overall level of play in each historical era. Those four shortstops were excluded from the normalization process in table 9.4 because, as noted above, the conditions prevailing when they played were not comparable to those of later eras.

The exclusion of these shortstops from table 9.4 does not mean that they were not great-hitting shortstops; it merely means that there is no way to determine just how great they were compared with the great-hitting shortstops who played later on. Nor does the inclusion of the four in table 9.3 constitute an estimate of their all-time ranking; it merely constitutes an estimate of their potential. There is no way to prove that potential, that is, to show how they would have performed if the conditions had been closer to those that prevailed in subsequent eras or if they had actually played in one of the subsequent eras.

The biggest gainer from the normalization process is Barry Larkin, who gained eight positions in the ranking. Lou Boudreau gained seven places, Robin Yount gained four places, and Honus Wagner and Ernie Banks gained three positions each through the normalization process. The biggest loser in the normalization process was Derek Jeter, who lost 10 positions. Miguel Tejada and Joe Sewell lost six positions each in the ranking because they played when league PRG averages were comparatively high. Honus Wagner replaces Alex Rodriguez as the best-hitting shortstop of all time, with Rodriguez a distant second and Ernie Banks third. Rodriguez is not likely to overtake Wagner, but even if he does, he will probably be classified as a third baseman by then, perhaps the best-hitting third baseman ever. Miguel Tejada and Derek Jeter could break into the top 10, but that remains to be seen. Otherwise, the order of the 10 greatest-hitting shortstops is likely to remain the same for the foreseeable future.

Twenty-two shortstops have been elected to the Baseball Hall of Fame, including 6 of the top 10 hitting shortstops in table 9.4. Wagner, Cronin, Banks, and Vaughan were probably elected based primarily on their hitting. Boudreau and Yount were also outstanding fielders and were probably elected based on a combination of their hitting and fielding. Ernie Banks was elected to the Baseball Hall of Fame, but as a first baseman. He has been included among the shortstops here because he was in his prime when he played shortstop and played nearly as many games there (1,125) as at first base (1,259).

Seven Hall of Fame shortstops not in table 9.4—Luis Aparacio, Rabbit Maranville, Phil Rizzuto, Ozzie Smith, Joe Tinker, Bobby Wallace, and John Ward—were undoubtedly elected because of their fielding accomplishments. They were all known for their outstanding defensive play and had anemic batting statistics—a combined batting average of .266, on-base percentage of .324, and slugging average of .345. Four other Hall of Fame shortstops not in table 9.4—Luke Appling, Dave Bancroft, Travis Jackson, and Pee Wee Reese—had higher batting statistics and were probably elected based primarily on their hitting. George Davis and Hugh Jennings were particularly known for their fielding prowess and were generally recognized as the leading all-around shortstops of their day. Joe Sewell had batting statistics comparable to Luke Appling, but he was also an excellent infielder. John Ward had weak batting statistics and was elected based on his other accomplishments. Cal Ripken was a good hitter and fielder but was probably elected based mainly on his remarkable streak of having played in 2,632 consecutive games. Thus, of the 22 shortstops elected to the Hall of Fame, 8 were probably elected based primarily on their hitting, 7 were probably elected based primarily on their fielding, and 7 were probably elected based on a combination of factors.

Second Basemen

The above comments about the size, speed, and power of shortstops also apply to second basemen. Because second basemen have to react quickly to a variety of critical situations, they tend to be quicker, smaller, and less productive hitters than most other position players. Table 9.1 indicates that second basemen tend to be the smallest or next to the smallest position players on the field. From 1876 to 1941 the leading second basemen averaged 5 feet 10 inches and 171 pounds, from 1942 to 2006 they averaged 6 feet and 179 pounds, and in 2007 they averaged 5 feet 11 inches and 192 pounds. Table 9.1 also indicates that second basemen are the third-fastest players on the field, ranking third in the number of bases stolen per game for all three periods of time. Table 9.2 indicates that second basemen, like shortstops, are not big hitters. Only one second baseman is among the top 50 hitters of all time in any of the weighted measures. There are six second basemen in the top 50 in OBP and four in the top 50 in AVG, but these are the two least reliable average measures of hitting.

The top 10 hitting second basemen of all time for each of the average measures of hitting performance are identified in table 9.5. The pattern of agreement and disagreement between measures is a familiar one—the same scene for all of the historical eras and for the position of shortstop. If you look at the top 10 as a group, there is general agreement on six second basemen who belong there. All of the measures agree on three (Hornsby, Gehringer, and Robinson), all except one measure agree on two more (Lajoie and Collins), and all except three measures agree on two others (Lazzeri and Childs). Twenty-one second basemen share the top 10 rankings, and 13 players share the top five rankings. Thus, there is a little less disagreement on the top 10 second basemen as a group than there was on the top 10 shortstops as a group.

If you look at the specific ranking of the second basemen within the top 10, there is more disagreement. Rogers Hornsby is ranked number 1 by all measures, a feat equaled by only two other position players: right fielder Babe Ruth and left fielder Ted Williams. Four measures agree that Gehringer should be number 4, Robinson number 5, and Lajoie number 7. None of the 17 remaining second basemen have more than two measures that agree on their specific ranking.

Table 9.6 summarizes the normalized PRG ratings for the top 10 hitting second basemen of all time. Two second basemen from the first two historical eras—Cupid Childs and Hardy Richardson— have not been included for the same reasons given in the preceding section on shortstops.

The biggest gainers from normalization were Eddie Collins, Jackie Robinson, and Joe Morgan, who improved three positions each in the rankings because they played when league PRG averages were low. Nap Lajoie gained two positions for the same reason. The biggest losers from normalization were Alfonso Soriano, who lost six positions, and Charlie Gehringer, who lost five positions, because they played when league PRG averages were high. Tony Lazzeri lost two positions in the rankings through normalization. There was no change in the ranking of Rogers Hornsby, Bobby Doerr, and Jeff Kent.

TABLE 9.5 The Top 10 Hitting Second Basemen of All Time According to Various Measures

Player	PRG	Advanced Weighted Measures			Basic Weighted Measures			Unweighted Measures	
		LSLR	RC/27	LWTS	TA	OPS	SLG	OBP	AVG
Rogers Hornsby	1	1	1	1	1	1	1	1	1
Tony Lazzeri	2	10	—	—	7	6	6	10	—
Charlie Gehringer	3	3	4	4	5	2	4	6	4
Nap Lajoie	4	—	7	7	8	7	7	9	2
Jeff Kent	5	—	—	10	—	4	3	—	—
Jackie Robinson	6	5	5	2	3	3	5	5	6
Bobby Doerr	7	—	—	—	—	9	9	—	—
Joe Gordon	8	—	—	—	—	10	8	—	—
Hardy Richardson	9	7	3	—	—	—	—	—	—
Eddie Collins	10	8	6	3	2	5	—	2	3
Cupid Childs	(11)	4	2	5	6	—	—	4	7
Alfonso Soriano	(12)	2	—	9	10	8	2	—	—
Joe Morgan	(15)	6	9	6	4	—	—	7	—
Roberto Alomar	(16)	9	—	8	—	—	—	—	10
Buddy Myer	(19)	—	—	—	—	—	—	8	9
Max Bishop	(29)	—	—	—	9	—	—	3	—
Other players[1]	—	—	—	—	—	—	—	—	—

[1] Five other second basemen were ranked in the top 10 but in one measure only: Bid McPhee #8 and Tom Daly #10 in RC/27, Ryne Sandberg #10 in SLG, and Frankie Frisch #5 and Billy Herman #8 in AVG.

Rogers Hornsby is undoubtedly the greatest-hitting second baseman ever—he is number 1 in PRG both before and after normalization and is first in every single one of the other measures we have been considering. Nap Lajoie is secure at number 2 with a substantial lead over the next two players closest to him. Alfonso Soriano is the only active player on the list, but he has been switched to left field and in a few years will have more games in left field than at second base. Thus, the order of the top 10 greatest-hitting second basemen is likely to remain the same for the foreseeable future.

Eighteen second basemen have been elected to the Baseball Hall of Fame. Eight of the nine second basemen in table 9.6 were elected

TABLE 9.6 Normalizing the PRG Ratings of the Top 10 Hitting Second Basemen

Before Normalization			Normalization		After Normalization		
			LG				
Rank	Rating	Player	PRG	NORM[1]	Rank	Rating	Player
1	1.283	Rogers Hornsby	.890	1.442	1	1.442	Rogers Hornsby*
2	1.156	Tony Lazzeri	.959	1.205	2	1.393	Nap Lajoie*
3	1.144	Charlie Gehringer	.964	1.187	3	1.209	Jackie Robinson*
4	1.128	Nap Lajoie	.810	1.393	4	1.205	Tony Lazzeri*
5	1.128	Jeff Kent	.936	1.205	5	1.205	Jeff Kent
6	1.104	Jackie Robinson	.913	1.209	6	1.201	Eddie Collins*
7	1.104	Bobby Doerr	.925	1.194	7	1.194	Bobby Doerr*
8	1.084	Joe Gordon	.928	1.168	8	1.187	Charlie Gehringer*
9	1.044	Eddie Collins	.869	1.201	9	1.168	Joe Gordon*
10	1.028	Alfonso Soriano	.958	1.073	10	1.167	Joe Morgan*[2]

* Indicates that a player was elected to the Baseball Hall of Fame.

[1] Player's PRG rating divided by his league's PRG rating for the years in which he played.

[2] Before normalization Morgan was ranked 13th with a PRG rating of 1.011 vs. a league average of .866.

based primarily on their hitting. Fielding was a factor in Joe Gordon's election, and Jeff Kent is not yet eligible because he retired in 2008. Five of the eight Hall of Famers not in table 9.6—Johnny Evers, Nellie Fox, Bill Mazeroski, Bid McPhee, and Red Schoendienst—were outstanding fielders. Since they all ranked low in hitting, they were probably chosen because of their fielding accomplishments. The three other Hall of Famers not in table 9.6—Frankie Frisch, Billy Herman, and Ryne Sandberg—had much better batting statistics and were good fielders. Hitting and fielding may have been equal considerations in their selection to the Hall of Fame. Rod Carew was undoubtedly elected based on his hitting—he was an outstanding batsman but a weak fielder. Thus, of the 18 second basemen elected to the Hall of Fame, 9 were probably elected based primarily on their hitting, and 9 were probably elected based primarily on their fielding, or on their hitting and fielding combined.

Third Basemen

When it comes to size and speed, third basemen rank in the middle of position players. They are ranked fourth in size and fifth or sixth in speed for the three periods of time covered in table 9.1. From 1876 to 1941 the leading third basemen averaged 5 feet 10 inches and 173 pounds, from 1942 to 2006 they averaged 6 feet 1 inch and 195 pounds, and in 2007 they averaged 6 feet 1 inch and 211 pounds. Table 9.1 also indicates that third basemen were somewhere in the middle of positions when it comes to speed—they rank fifth or sixth in stolen bases per game in each period of time. Among infielders only, third basemen also tend to be ranked in the middle—bigger and slower than shortstops and second basemen but smaller and faster than first basemen. Third basemen also tend to be somewhere in the middle when it comes to hitting. Outfielders and first basemen are clearly the leading hitters in table 9.2. All other position players are far behind, but third basemen do tend to lead second basemen, shortstops, and catchers in all measures except the two least reliable ones, on-base percentage and batting average.

The top 10 hitting third basemen of all time according to the various measures of hitting performance are identified in table 9.7. The pattern of agreement and disagreement between measures is no different from what we have seen before. If you look at the top 10 as a group, there is general agreement on eight who belong in the group. All of the measures agree on only one (Jones), but all except one measure agree on two others (Schmidt and Mathews), and all except two measures agree on four more (Joyce, Rosen, Lyons, and McGraw). Twenty-five third basemen share the top 10 rankings, and 13 share the top five rankings. Thus, there is less disagreement on the top 10 third basemen as a group than on second basemen and shortstops as a group. If you look at the specific ranking of the players within the top 10, there is more disagreement. McGraw is ranked number 1 by five measures, and Joyce is ranked number 2 and Jones number 3 by four measures each. Lyons is ranked number 5 and Rosen and Rolen number 7 by three measures each. None of the remaining 19 third basemen have more than two measures that agree on their specific ranking.

Table 9.8 summarizes the normalized PRG ratings for the top 10 hitting third basemen of all time. Six players from the first two historical eras (Bill Joyce, Denny Lyons, John McGraw, Deacon White,

TABLE 9.7 The Top 10 Hitting Third Basemen of All Time According to Various Measures

Player	PRG	Advanced Weighted Measures			Basic Weighted Measures			Unweighted Measures	
		LSLR	RC/27	LWTS	TA	OPS	SLG	OBP	AVG
Bill Joyce	1	1	2	2	2	3	—	2	—
Miguel Cabrera	2	—	5	—	—	—	—	—	—
Chipper Jones	3	3	4	3	3	1	1	5	6
Mike Schmidt	4	5	8	6	4	2	2	8	—
Al Rosen	5	7	—	8	8	5	7	7	—
Denny Lyons	6	4	3	5	5	—	—	4	5
Scott Rolen	7	10	—	9	7	7	5	—	—
Aramis Ramirez	8	—	—	—	—	—	4	—	—
Troy Glaus	9	—	—	—	9	10	6	—	—
Eddie Mathews	10	6	9	7	6	4	3	9	—
George Brett	(11)	9	—	10	—	9	9	—	9
John McGraw	(15)	2	1	1	1	6	—	1	1
Harlond Clift	(22)	—	—	—	10	—	—	6	—
Wade Boggs	(25)	8	—	4	—	8	—	3	2
Other players[1]	—	—	—	—	—	—	—	—	—

[1] Eleven other third basemen were ranked in the top 10, but in one measure only: Deacon White #6, Billy Nash #7, and Ned Williamson #10 in RC/27; Matt Williams #8 and Eric Chavez #10 in SLG; Bob Elliott #10 in OBP; and Pie Traynor #3, Freddie Lindstrom #4, Home Run Baker #7, George Kell #8, and Bill Madlock #10 in AVG.

Billy Nash, and Ned Williamson) were not included. The biggest gainers from normalization were Home Run Baker and Ron Santo, who gained eight places each in the rankings, because they played when league PRG averages were low. Troy Glaus was the biggest loser from normalization: he lost eight positions because he played when league PRG averages were high. Pie Traynor and Aramis Ramirez lost six positions each and Scott Rolen five positions from normalization for the same reason. The rankings of Mike Schmidt, Miguel Cabrera, Al Rosen, Bob Elliott, and George Brett changed only minimally, and the ranking of Chipper Jones was unchanged.

The rankings of the top 10 hitting third basemen will probably change significantly in the years ahead. Alex Rodriguez will have

TABLE 9.8 Normalizing the PRG Ratings of the Top 10 Hitting Third Basemen

Before Normalization			Normalization		After Normalization		
Rank	Rating	Player	LG PRG	NORM[1]	Rank	Rating	Player
1	1.205	Miguel Cabrera	.948	1.271	1	1.338	Mike Schmidt*
2	1.203	Chipper Jones	.939	1.281	2	1.281	Chipper Jones
3	1.165	Mike Schmidt	.871	1.338	3	1.271	Miguel Cabrera
4	1.164	Al Rosen	.920	1.265	4	1.265	Al Rosen
5	1.140	Scott Rolen	.944	1.208	5	1.260	Eddie Mathews*
6	1.113	Aramis Ramirez	.938	1.187	6	1.259	Home Run Baker*[2]
7	1.113	Troy Glaus	.944	1.179	7	1.255	Ron Santo[3]
8	1.113	Eddie Mathews	.883	1.260	8	1.213	George Brett*
9	1.110	George Brett	.915	1.213	9	1.212	Bob Elliott[4]
10	1.109	Pie Traynor	.951	1.166	10	1.208	Scott Rolen

* Indicates that a player was elected to the Baseball Hall of Fame.

[1] Player's PRG rating divided by his league's PRG rating for the years in which he played.

[2] Before normalization Baker was ranked 14th with a PRG rating of 1.075 vs. a league average of .854.

[3] Before normalization Santo was ranked 15th with a PRG rating of 1.073 vs. a league average of .855.

[4] Before normalization Elliott was ranked 11th with a PRG rating of 1.098 vs. a league average of .906.

played more games at third base than at shortstop and will be classi-fied as a third baseman. Since he is only 35 years old, his 1.266 PRG rating (after normalization) may improve, putting him near the top of the third basemen, depending on what the 38-year-old Chipper Jones does in the meantime. Jones suffered from injuries that have prevented him from playing a full schedule in recent years. Thus, his rating could decline, at least somewhat, before he retires from base-ball. One other active third baseman will also be a contender for a higher ranking. Scott Rolen is 35 years old and could, with some improvement in his statistics, advance in the rankings. Miguel Ca-brera has played more games at third base than at any other position but has been moved to first base the last two years and soon will have more games at that position.

Only 10 third basemen have been elected to the Baseball Hall of Fame, the fewest of any position. Four of the top 10 hitting third basemen in table 9.8 are in the Hall of Fame. Three others, Chipper Jones, Scott Rolen, and Miguel Cabrera, are not eligible because they are active players. One of the Hall of Famers not in table 9.8, Freddie Lindstrom, was good both at bat and in the field and was probably elected based on a combination of both factors. Two others—Jimmy Collins and Brooks Robinson—were undoubtedly elected because of their legendary fielding accomplishments. George Kell—a very good, but not great, hitter and fielder—was probably elected based on a combination of his hitting and fielding. Pie Traynor was a good hitter and was probably elected based on his hitting. Wade Boggs learned to play a very good third base, but it was hitting that brought his election to the Hall of Fame. His overall PRG rating is low because he was mainly a leadoff batter who had fewer opportunities to advance runners and drive in runs, key elements in the PRG formula. His performance as a leadoff batter will be covered in the next chapter. Thus, of the 10 third basemen elected to the Hall of Fame, 6 were probably elected based primarily on their hitting, 2 were probably elected based primarily on their fielding, and 2 were probably elected based on their hitting and fielding combined.

First Basemen

Fielding is less important in relation to hitting for first basemen than for any other position on the field. First basemen tend, therefore, to be big and slow, but good hitters. Table 9.1 indicates that first basemen are, on average, the biggest and slowest position players both in the past and in the present. From 1876 to 1941 the leading first basemen averaged 6 feet and 194 pounds, from 1942 to 2006 they averaged 6 feet 2 inches and 208 pounds, and in 2007 they averaged 6 feet 3 inches and 225 pounds. For each period of time, they are the biggest and second-slowest position players of all. They have, however, been surpassed in both size and slowness by American League designated hitters. Table 9.2 indicates that first basemen, on average, are the best-hitting players on the field. For seven of the nine measures, there are more first basemen among the top 50 hitters of all time than for any other position. First basemen rank second in the RC/27 model and are tied for third in the AVG model.

The top 10 first basemen of all time according to the various measures of hitting performance are identified in table 9.9. If you look

TABLE 9.9 The Top 10 Hitting First Basemen of All Time According to Various Measures

Player	PRG	Advanced Weighted Measures			Basic Weighted Measures			Unweighted Measures	
		LSLR	RC/27	LWTS	TA	OPS	SLG	OBP	AVG
Lou Gehrig	1	1	2	1	1	1	1	1	3
Hank Greenberg	2	3	6	4	4	4	4	7	10
Jimmie Foxx	3	5	3	3	2	3	3	2	8
Albert Pujols	4	2	7	2	3	2	2	4	4
Dan Brouthers	5	4	1	5	7	—	—	6	1
Lance Berkman	6	6	—	7	8	7	9	8	—
Johnny Mize	7	—	10	—	—	9	7	—	—
Mark McGwire	8	9	8	9	6	6	5	—	—
Todd Helton	9	7	9	6	5	5	6	3	6
Jim Thome	10	10	—	10	9	8	8	—	—
Jeff Bagwell	(13)	8	—	8	10	10	—	9	—
Cap Anson	(15)	—	5	—	—	—	—	—	5
Roger Connor	(18)	—	4	—	—	—	—	—	9
Other players[1]	—	—	—	—	—	—	—	—	—

[1] Four other first basemen were ranked in the top 10 but in one measure only: Carlos Delgado #10 in SLG, Ferris Fain #5 in OBP, and Bill Terry #2 and Rod Carew #7 in AVG.

at the top 10 first basemen as a group, there is general agreement on nine who belong in that group. All of the measures agree on five (Gehrig, Greenberg, Foxx, Pujols, and Helton), all except two measures agree on three (McGwire, Berkman, and Brouthers), and all except three measures agree on one more (Thome). Only 17 first basemen (the least for any position) share the top 10 rankings, and only 11 share the top five rankings. Thus, there is less disagreement on the top 10 first basemen as a group than on any other position.

If you look at the specific ranking of first basemen within the top 10, there is more disagreement but not as much as there is for other positions. Gehrig is ranked number 1 by seven measures, and Foxx is ranked number 3 by five measures. Pujols is ranked number 2 and Greenberg number 4 by four measures each, and two others are ranked the same by three different measures. None of the remaining

11 first basemen have more than two measures that agree on their specific ranking. There is less disagreement on the specific ranking of first basemen than any other position.

Table 9.10 summarizes the normalized PRG ratings for the top 10 hitting first basemen of all time. In keeping with the procedure with regard to the other infield positions, three first basemen from the Era of Constant Change—Dan Brouthers, Cap Anson, and Roger Connor—have not been included for the reasons previously given. The difference between the first baseman with the most plate appearances (Harmon Killebrew—9,831) and the first baseman with the least plate appearances (Hank Greenberg—6,096) is much smaller with regard to first basemen than for the other infield positions. Hank Greenberg is generally acclaimed as one of the greatest first basemen ever and would have had thousands more plate appearances had he not lost nearly five years of baseball because of his service to his country in World War II.

The big gainers from normalization were Willie McCovey and Harmon Killebrew, who improved 12 places each in the rankings because they played when league PRG averages were low. Dick Allen gained 11 places in the rankings for the same reason. The big losers from normalization were Jim Thome and Carlos Delgado, who lost six and seven places, respectively, because they are playing when league PRG averages are high. Todd Helton lost four places for the same reason. The PRG rankings of the top two first basemen— Gehrig and Greenberg—remain the same. Johnny Mize and Albert Pujols gained one place while Jimmie Foxx and Mark McGwire lost one place each. Lance Berkman lost two places.

Lou Gehrig is undoubtedly the greatest-hitting first baseman ever—he is first in PRG both before and after normalization and is first in six of the eight other measures we have been considering. Albert Pujols is the highest-ranking active first baseman on the normalized list (#3). He is only 30 years old and could improve his batting statistics enough to overtake Hank Greenberg and end up in second place. The only other active first baseman on the top 10 list is Lance Berkman (#7). Berkman is 34 years old—his statistics could either improve or decline in the next few years. He could improve

TABLE 9.10 Normalizing the PRG Ratings of the Top 10 Hitting First Basemen

Before Normalization			Normalization		After Normalization		
Rank	Rating	Player	LG PRG	NORM[1]	Rank	Rating	Player
1	1.429	Lou Gehrig	.966	1.479	1	1.479	Lou Gehrig*
2	1.381	Hank Greenberg	.954	1.448	2	1.448	Hank Greenberg*
3	1.366	Jimmie Foxx	.959	1.424	3	1.427	Albert Pujols
4	1.333	Albert Pujols	.934	1.427	4	1.424	Jimmie Foxx*
5	1.269	Lance Berkman	.939	1.351	5	1.406	Johnny Mize*
6	1.265	Johnny Mize	.900	1.406	6	1.360	Dick Allen[2]
7	1.265	Mark McGwire	.946	1.337	7	1.351	Lance Berkman
8	1.251	Todd Helton	.950	1.317	8	1.337	Mark McGwire
9	1.236	Jim Thome	.976	1.266	9	1.337	Willie McCovey*[3]
10	1.230	Carlos Delgado	.974	1.263	10	1.333	Harmon Killebrew*[4]

* Indicates that a player was elected to the Baseball Hall of Fame.

[1] Player's PRG rating divided by his league's PRG rating for the years in which he played.

[2] Before normalization Allen was ranked 17th with a PRG rating of 1.156 vs. a league average of .850.

[3] Before normalization McCovey was ranked 21st with a PRG rating of 1.151 vs. a league average of .861.

[4] Before normalization Killebrew was ranked 22nd with a PRG rating of 1.148 vs. a league average of .861.

enough to overtake Dick Allen (#6) or decline enough to fall behind Mark McGwire (#8) or even Willie McCovey (#9) and Harmon Killebrew (#10). Todd Helton is the next highest-ranking active first baseman not on the top 10 list (#12). He is 36 years old, but with a couple of good years he could break into the top 10.

Twenty first basemen have been elected to the Baseball Hall of Fame, including Ernie Banks and Rod Carew, who are treated here as a shortstop and second baseman, respectively. Six of the remaining 18 were top hitting first basemen included in table 9.10. One player in that table—Mark McGwire—would have likely been elected had he not been tarnished because of his role in the current steroids controversy. A second player—Dick Allen—had the batting statistics to be elected but was probably rejected because he was constantly involved in personal controversies of one kind or another.

Most of the remaining 12 Hall of Fame first basemen not in table 9.10 were also good hitters. In fact, it's hard to imagine that any first baseman, except perhaps Frank Chance, was elected to the Hall of

Fame based primarily on his fielding. Chance did not have an impressive batting resumé, but he was the first baseman in the famous Tinker-to-Evers-to-Chance double-play combination. All three of these players were elected to the Hall of Fame together in 1946. Thus, fielding versus hitting is much less important for first basemen than for any other position.

Catchers

The physical demands on catchers are the most for any position player—their hitting has to be weighed not only against their fielding but also against their handling of pitchers. And a catcher's fielding has important elements that cannot be compromised without great risk to the team. Catchers must be physical so they can block the plate against runners trying to score, they must have strong arms so they can throw out runners attempting to steal, and they must have good hands so they can block errant throws from the pitcher.

Catchers, therefore, tend to be bigger and slower and not among the top hitters on teams. Table 9.1 indicates that catchers are the second-biggest position players on the field in both the past and the present. From 1876 to 1941 the leading catchers averaged 5 feet 11 inches and 187 pounds, from 1942 to 2006 they averaged 6 feet and 207 pounds, and in 2007 they averaged 6 feet 1 inch and 213 pounds. Table 9.1 also indicates that catchers are the slowest position players. In each of the three periods of time, catchers stole the fewest bases per game. Catchers tend to be big physically, but table 9.2 indicates that they are not big hitters. Only one catcher at most is ranked among the top 50 hitters of all time in any of the various measures.

The top 10 hitting catchers of all time according to the various measures of hitting performance are identified in table 9.11. The pattern of agreement and disagreement between the measures is similar to what we have seen for the infield positions. If you look at the top 10 catchers as a group, there is general agreement on six who belong in the group. All of the measures agree on three catchers (Piazza, Dickey, and Cochrane), all except one measure agree on two other catchers (Posada and Hartnett), and all except two measures agree

TABLE 9.11 The Top 10 Hitting Catchers of All Time According to Various Measures

| Player | PRG | Advanced Weighted Measures | | | Basic Weighted Measures | | | Unweighted Measures | |
		LSLR	RC/27	LWTS	TA	OPS	SLG	OBP	AVG
Mike Piazza	1	1	4	1	2	1	1	8	4
Bill Dickey	2	5	3	5	5	3	5	5	2
Roy Campanella	3	4	8	6	8	4	2	—	—
Jorge Posada	4	6	7	4	4	6	8	6	—
Gabby Hartnett	5	—	6	7	7	5	4	9	9
Mickey Cochrane	6	3	2	2	1	2	7	1	1
Yogi Berra	7	7	9	9	—	7	6	—	—
Buck Ewing	8	2	1	3	6	—	—	—	6
Ernie Lombardi	9	—	—	—	—	9	—	—	5
Johnny Bench	10	8	—	—	—	—	9	—	—
Javy Lopez	(12)	—	—	—	—	8	3	—	—
Gene Tenace	(13)	—	10	10	3	—	—	3	—
Mickey Tettleton	(14)	—	—	8	9	10	—	—	—
Spud Davis	(16)	—	—	—	—	—	—	10	3
Joe Torre	(17)	10	—	—	—	—	—	—	8
Ivan Rodriguez	(22)	9	—	—	—	—	10	—	7
Wally Schang	(25)	—	—	—	10	—	—	2	—
Other players[1]	—	—	—	—	—	—	—	—	—

[1] Four other catchers were ranked in the top 10 but in one measure only: Jack Clements #5 in RC/27, Roger Bresnahan #4 and Rick Ferrell #7 in OBP, and Manny Sanguillen #10 in AVG.

on another (Campanella). Twenty-one catchers share the top 10 rankings, and 14 catchers share the top five rankings.

If you look at the specific rankings of the catchers within the top 10, there is more disagreement. Five measures agree that Piazza should be ranked number 1 and that Dickey should be ranked number 5, whereas three measures agree that Cochrane should be ranked number 1, Posada should be ranked number 6, and Berra should be ranked number 7. None of the remaining 16 catchers have more than two measures that agree on their specific ranking.

Table 9.12 summarizes the normalized PRG ratings and rankings

of the top 10 catchers of all time. Two catchers from table 9.11—
Buck Ewing and Jack Clements—are not included because they
played in the Era of Constant Change.

The biggest gainer by far from normalization was Roger Bresnahan, who gained 15 places in the rankings because he played in the
Dead Ball Era when league averages were at their very lowest. Other
significant gainers from normalization are Joe Torre, who gained
seven places, and Johnny Bench, who gained six places in the rankings. Roy Campanella, Yogi Berra, and Gene Tenace also gained two
places each from normalization. The biggest losers from normalization were Jorge Posada, who lost 11 places, Mickey Cochrane, who
lost seven, and Bill Dickey, who lost four. Mike Piazza recently retired, making Jorge Posada the highest-ranking active catcher. Posada had a great year in 2007, missed most of the 2008 season because
of injuries, and had a good year in 2009. He could climb back into
the top 10, but he is 39 years old and time is running out. Otherwise,
the all-time ranking of catchers is likely to remain the same.

Thirteen catchers have been elected to the Baseball Hall of Fame,
including 7 of the top 10 hitting catchers in table 9.12. Roger Bresnahan was one of the leading all-around players of his day and was
probably elected based on a combination of his hitting and fielding.
The other six were also good defensive catchers, especially Bench and
Hartnett, but their hitting was so great for catchers that they probably would have been elected had they been only average-fielding
catchers. Joe Torre and Gene Tenace have never been elected, and
Mike Piazza is not yet eligible because he just retired.

One other catcher—Buck Ewing—was also elected based primarily on his hitting. He played at a time when catchers split their time
between catching and other positions and played only 48% of his
games as a catcher. Ewing was, however, the best-hitting catcher of
his time, and many thought that he was the best player of his day
regardless of position. Gary Carter, Rick Ferrell, and Ray Schalk
were not great hitters and were undoubtedly elected based largely on
their fielding accomplishments. Mickey Cochrane was a good hitter
and fielder and was probably elected based on both factors. Carlton
Fisk was a very durable slugger. Until recently he held the all-time

TABLE 9.12 Normalizing the PRG Ratings of the Top 10 Hitting Catchers

	Before Normalization		Normalization			After Normalization	
Rank	Rating	Player	LG PRG	NORM[1]	Rank	Rating	Player
1	1.188	Mike Piazza	.938	1.267	1	1.274	Roy Campanella*
2	1.179	Bill Dickey	.958	1.231	2	1.267	Mike Piazza
3	1.159	Roy Campanella	.910	1.274	3	1.256	Johnny Bench*
4	1.149	Jorge Posada	.983	1.169	4	1.238	Gabby Hartnett*
5	1.140	Gabby Hartnett	.921	1.238	5	1.235	Yogi Berra*
6	1.138	Mickey Cochrane	.964	1.180	6	1.231	Bill Dickey*
7	1.121	Yogi Berra	.908	1.235	7	1.225	Joe Torre[2]
8	1.089	Ernie Lombardi	.893	1.219	8	1.219	Ernie Lombardi*
9	1.084	Johnny Bench	.863	1.256	9	1.211	Roger Bresnahan*[3]
10	1.069	Walker Cooper	.893	1.197	10	1.207	Gene Tenace[4]

* Indicates that a player was elected to the Baseball Hall of Fame.
[1] Player's PRG rating divided by his league's PRG rating for the years in which he played.
[2] Before normalization Torre was ranked 14th with a PRG rating of 1.049 vs. a league average of .856.
[3] Before normalization Bresnahan was ranked 24th with a PRG rating of .974 vs. a league average of .804.
[4] Before normalization Tenace was ranked 12th with a PRG rating of 1.055 vs. a league average of .874.

record for games caught by a catcher and still ranks third in home runs (376) for a catcher. Thus, of the 13 catchers elected to the Hall of Fame, 7 were probably elected based primarily on their hitting, 3 were probably elected based primarily on their fielding, and 3 were probably elected based on their hitting and fielding combined.

Center Fielders

Center fielders have to be faster than their partners on either side of them in the outfield because they have more ground to cover and more balls to chase down. Since size is inversely related to speed (the smaller they are, the faster they are), center fielders also tend to be smaller than left fielders and right fielders. Table 9.1 confirms both of these statements. Center fielders were the fastest players, not only in the outfield but on the entire field, in all three of the periods of time covered in the table. From 1876 to 1941 center fielders stole a base in 15% of their games (vs. 12% for left fielders and right fielders), from 1942 to 2006 in 10% of their games (vs. 7% for left fielders and right fielders), and in 2007 in 14% of their games (vs. 6% for left fielders and right fielders).

Center fielders, on average, were also smaller than left fielders and right fielders for all three periods of time. From 1876 to 1941 center fielders averaged 5 feet 10 inches and 174 pounds (vs. 5 feet 10.5 inches and 177 pounds for left fielders and right fielders), from 1942 to 2006 they averaged 6 feet and 188 pounds (vs. 6 feet 1.5 inches and 196.5 pounds for left fielders and right fielders), and in 2007 they averaged 6 feet and 196 pounds (vs. 6 feet 1.5 inches and 210.5 pounds for left fielders and right fielders). Since size is positively related to power, left and right fielders tend to be more productive hitters than center fielders. Thus, most of the hitting models in table 9.2 include fewer center fielders than left or right fielders on their lists of the top 50 hitters of all time. The PRG model includes only six center fielders versus 11 left fielders and 10 right fielders among the top 50 hitters of all time. Only the RC/27 and TA models list more center fielders, and then only slightly more, than left or right fielders.

The top 10 hitting center fielders of all time according to the various measures of hitting performance are identified in table 9.13. The pattern of agreement and disagreement persists. If you look at the top 10 center fielders as a group, there is general agreement on only five who belong in that group (tied with right fielders for the fewest for any position). All of the measures agree on two center fielders (DiMaggio and Cobb), all except one agree on another two (Mantle and Speaker), and all except two agree on one more (Averill). Twenty-one center fielders share the top 10 rankings, and 11 center fielders share the top five rankings.

If you look at the specific rankings of the center fielders within the top 10, there is more disagreement. Five measures agree that Hamilton is number 1, and four measures agree that Cobb is number 3 and Mantle is number 4. None of the 18 remaining center fielders have more than two measures that agree on their specific ranking.

The rankings in table 9.13 are interesting, but they are not definitive, because they include players from the Era of Constant Change (Pete Browning, George Gore, and Bug Holliday) and Live Ball Interval (Hugh Duffy, Billy Hamilton, and Mike Griffin). Table 9.14

TABLE 9.13 The Top 10 Hitting Center Fielders of All Time According to Various Measures

Player	PRG	Advanced Weighted Measures			Basic Weighted Measures			Unweighted Measures	
		LSLR	RC/27	LWTS	TA	OPS	SLG	OBP	AVG
Joe DiMaggio	1	2	4	3	5	2	1	6	8
Hack Wilson	2	—	—	8	6	5	4	9	—
Ty Cobb	3	3	5	2	3	3	10	2	1
Mickey Mantle	4	4	3	4	2	1	3	4	—
Earl Averill	5	6	—	9	8	6	7	8	—
Duke Snider	6	—	—	—	—	8	6	—	—
Ken Griffey Jr.	7	8	—	—	9	9	5	—	—
Willie Mays	8	5	—	6	7	4	2	—	—
Tris Speaker	9	10	8	5	4	7	—	3	2
Jim Edmonds	10	—	—	—	—	10	8	—	—
Hugh Duffy	(11)	9	7	10	—	—	—	—	6
Pete Browning	(16)	7	2	7	10	—	—	5	4
Billy Hamilton	(19)	1	1	1	1	—	—	1	3
Earle Combs	(28)	—	—	—	—	—	—	7	7
Mike Griffin	(32)	—	9	—	—	—	—	10	—
Other players[1]	—	—	—	—	—	—	—	—	—

[1] Six other center fielders were ranked in the top 10 but in one measure only: George Gore #6 and Bug Holliday #10 in RC/27; Wally Berger #9 in SLG; and Mike Donlin #5, Edd Roush #9, and Kirby Puckett #10 in AVG.

excludes those center fielders. The biggest gainer from normalization was Mike Donlin, who gained 10 places because he played in the Dead Ball Era when league PRG averages were at their lowest. Tris Speaker gained five places for the same reason. Willie Mays gained three places, and Larry Doby and Ty Cobb gained two places each. Earl Averill and Ken Griffey Jr. lost five places each. Hack Wilson lost four places and Jim Edmonds lost three places because they played when PRG averages were high. All other changes are minimal. No active players are on the top 10 list.

Seventeen center fielders have been elected to the Baseball Hall of Fame, including 9 of the top 10 hitting center fielders identified in table 9.14. Two center fielders not in table 9.14 were also elected

TABLE 9.14 Normalizing the PRG Ratings of the Top 10 Hitting Center Fielders

Before Normalization			Normalization		After Normalization		
Rank	Rating	Player	LG PRG	NORM[1]	Rank	Rating	Player
1	1.326	Joe DiMaggio	.941	1.409	1	1.428	Ty Cobb*
2	1.269	Hack Wilson	.955	1.329	2	1.409	Joe DiMaggio*
3	1.215	Ty Cobb	.851	1.428	3	1.386	Mickey Mantle*
4	1.214	Mickey Mantle	.876	1.386	4	1.345	Tris Speaker*
5	1.213	Earl Averill	.973	1.247	5	1.338	Willie Mays*
6	1.178	Duke Snider	.900	1.309	6	1.329	Hack Wilson*
7	1.173	Ken Griffey Jr.	.954	1.228	7	1.309	Duke Snider*
8	1.169	Willie Mays	.874	1.338	8	1.297	Mike Donlin[2]
9	1.163	Tris Speaker	.865	1.345	9	1.256	Larry Doby*[3]
10	1.158	Jim Edmonds	.960	1.206	10	1.247	Earl Averill*

* Indicates that a player was elected to the Baseball Hall of Fame.

[1] Player's PRG rating divided by his league's PRG rating for the years in which he played.

[2] Before normalization Donlin was ranked 18th with a PRG rating of 1.073 vs. a league average of .827.

[3] Before normalization Doby was ranked 11th with a PRG rating of 1.148 vs. a league average of .914.

to the Hall of Fame based primarily on their hitting. Billy Hamilton was the premier leadoff batter and one of the best hitters overall during his day. He was also an excellent base runner, but he was not known for his fielding. Hugh Duffy was known for his fielding, but his hitting was so good that he could have been elected based on his hitting alone. Max Carey's election was undoubtedly based on his fielding, since he was one of the best-fielding center fielders of his day and was a weak hitter. Five other center fielders in the Hall of Fame were probably elected based on a combination of their hitting and fielding. Edd Roush and Kirby Puckett were exceptional in the field and at the plate, and both factors probably were the basis for their election to the Hall of Fame. Richie Ashburn, Earle Combs, and Lloyd Waner were leadoff batters and were probably elected based on their all-around abilities because they were good hitters, fielders, and base runners. Leadoff batters will be discussed as a separate category of hitters in the next chapter.

In summary, then, 11 of the 17 Hall of Fame center fielders were probably elected based on their hitting—the nine in table 9.14 plus Hamilton and Duffy. One of the 17, Carey, was probably elected based on his fielding alone, and 5 of the 17 were probably elected based on a combination of their hitting and fielding—Ashburn, Combs, Puckett, Roush, and Waner.

Left Fielders

Since left fielders have less ground to cover and shorter throws to third base than center fielders, fielding is not as important versus hitting for left fielders as it is for center fielders. As a result, left fielders tend to be slower, bigger, and better sluggers than center fielders. Table 9.1 supports this statement. Left fielders were bigger and slower than center fielders in all three periods of time covered in the table. From 1876 to 1941 left fielders stole a base in 11% of their games (vs. 15% for center fielders), from 1942 to 2006 in 7% of their games (vs. 10% for center fielders), and in 2007 in 6% of their games (vs. 14% for center fielders). Left fielders, on average, were also bigger than center fielders for all three periods of time. From 1876 to 1941 left fielders averaged 5 feet 11 inches and 180 pounds (vs. 5 feet 10 inches and 174 pounds for center fielders), from 1942 to 2006 they averaged 6 feet 1 inch and 201 pounds (vs. 6 feet and 188 pounds for center fielders), and in 2007 they averaged 6 feet 1 inch and 209 pounds (vs. 6 feet and 196 pounds for center fielders). Since size is related to power (the bigger the player, the more power he has), left fielders also tend to be better sluggers than center fielders. Thus, all of the hitting models in table 9.2 except LWTS include more left fielders than center fielders on their lists of the top 50 hitters of all time. Overall, left fielders have the second-most players on the lists of six of the nine models and have the most players listed in the RC/27 model.

The top 10 hitting left fielders of all time according to the various measures of hitting performance are identified in table 9.15. If you look at the top 10 left fielders as a group, there is general agreement on six who belong in that group. All of the measures agree on two left fielders (Williams and Musial), and all except one measure agree on

TABLE 9.15 The Top 10 Hitting Left Fielders of All Time According to Various Measures

Player	PRG	Advanced Weighted Measures			Basic Weighted Measures			Unweighted Measures	
		LSLR	RC/27	LWTS	TA	OPS	SLG	OBP	AVG
Ted Williams	1	1	1	1	1	1	1	1	3
Manny Ramirez	2	4	9	3	3	3	3	7	—
Barry Bonds	3	2	3	2	2	2	2	2	—
Ed Delahanty	4	3	4	4	5	10	—	6	2
Al Simmons	5	—	—	—	—	—	7	—	6
Albert Belle	6	5	—	8	—	7	4	—	—
Stan Musial	7	7	7	5	4	4	5	4	7
Charlie Keller	8	—	10	—	7	8	—	8	—
Ralph Kiner	9	6	—	6	8	5	6	—	—
Joe Jackson	10	10	8	7	6	6	—	3	1
Tip O'Neill	(13)	—	2	—	—	—	—	—	9
Joe Kelley	(19)	—	—	10	10	—	—	10	—
Riggs Stephenson	(22)	—	—	—	—	—	—	9	5
Harry Stovey	(35)	9	5	—	—	—	—	—	—
Ken Williams	(43)	—	—	—	9	9	8	—	—
Jesse Burkett	(47)	8	6	9	—	—	—	5	4
Other players[1]	—	—	—	—	—	—	—	—	—

[1] Four other left fielders were ranked in the top 10 but in one measure only: Willie Stargell #9 and Chick Hafey #10 in SLG, and Heine Manush #8 and Joe Medwick #10 in AVG.

four more (Ramirez, Barry Bonds, Delahanty, and Jackson). Twenty left fielders share the top 10 rankings, and 13 of them share the top five rankings.

If you look at the specific rankings of the left fielders within the top 10, there is more disagreement. Eight measures agree that Williams is number 1, six measures agree that Bonds is number 2, and four measures agree that Ramirez is number 3 and Musial is number 7. Three measures agree on the specific ranking of Delahanty, Keller, Kiner, and Kelley. None of the remaining 12 left fielders have more than two measures that agree on their specific ranking.

The player rankings in table 9.15 are based on raw data. Table 9.16 deletes players who played in the Era of Constant Change (Tip

O'Neill and Harry Stovey) and Live Ball Interval (Ed Delahanty, Joe Kelley, and Jesse Burkett). The biggest gainers in table 9.16 from normalization were Willie Stargell (plus 10 positions) and Joe Jackson (plus seven positions). They played at times when conditions favored pitchers over batters, but they had good batting statistics, thus making their accomplishments greater than they would appear at first glance. If they had played when hitters dominated pitchers, their batting statistics probably would have been higher, and normalization adjusts their batting statistics to compensate for this. The biggest losers from normalization were Albert Belle (minus eight positions) and Al Simmons (minus six positions), who played when batters tended to dominate pitchers. Manny Ramirez and Bob Johnson also played during such times and lost five and four places, respectively, in the rankings. There was no change in the rankings of Ted Williams, Barry Bonds, and Ralph Kiner.

Manny Ramirez is the only active player among the top 10 left fielders of all time. He has a normalized PRG rating of 1.347, very close to Keller (1.350) and within striking distance of Stargell

TABLE 9.16 Normalizing the PRG Ratings of the Top 10 Hitting Left Fielders

Before Normalization			Normalization		After Normalization		
			LG				
Rank	Rating	Player	PRG	NORM[1]	Rank	Rating	Player
1	1.428	Ted Williams	.915	1.561	1	1.561	Ted Williams*
2	1.321	Manny Ramirez	.981	1.347	2	1.423	Joe Jackson
3	1.291	Barry Bonds	.915	1.411	3	1.411	Barry Bonds
4	1.253	Al Simmons	.963	1.301	4	1.380	Stan Musial*
5	1.245	Albert Belle	.976	1.276	5	1.364	Willie Stargell*[2]
6	1.234	Stan Musial	.894	1.380	6	1.350	Charlie Keller
7	1.230	Charlie Keller	.911	1.350	7	1.347	Manny Ramirez
8	1.203	Ralph Kiner	.909	1.323	8	1.323	Ralph Kiner*
9	1.190	Joe Jackson	.836	1.423	9	1.321	Joe Medwick*[3]
10	1.189	Bob Johnson	.934	1.273	10	1.301	Al Simmons*

* Indicates that a player was elected to the Baseball Hall of Fame.
[1] Player's PRG rating divided by his league's PRG rating for the years in which he played.
[2] Before normalization Stargell was ranked 15th with a PRG rating of 1.166 vs. a league average of .855.
[3] Before normalization Medwick was ranked 13th with a PRG rating of 1.174 vs. a league average of .889.

(1.364). He could overtake Stargell, but he is 38 years old and the odds appear to be against him.

Twenty left fielders, including the six identified in table 9.16, have been elected to the Baseball Hall of Fame. Ten of the other 14 Hall of Fame left fielders were also probably elected based primarily on their hitting. Three of them—Jesse Burkett, Ed Delahanty, and Heine Manush—had lifetime batting averages of .330 or more, and none of them were known for their fielding. Goose Goslin had a lifetime batting average of .317 and a slugging average of .500 and was not known for his fielding. Two other left fielders—Jim O'Rourke and Zach Wheat—had lifetime batting averages of .310 or more and were not known for their fielding. Jim Rice had a lifetime batting average of .298 and was not known for his fielding, but he averaged 27 home runs and 101 runs batted in and had a slugging average of .507 for 14 straight years. Billy Williams and Carl Yastrzemski had lifetime batting averages under .300, but they played during the Dead Ball Interval when pitchers had the upper hand on most batters. Yet Billy Williams had a slugging average of nearly .500 and Yastrzemski had three batting titles and a Triple Crown to his credit. Joe Kelley was undoubtedly elected based primarily on his performance as a hitter.

Two left fielders—Fred Clarke and Chick Hafey—were probably elected based on a combination of their hitting and fielding. Clarke had a lifetime batting average of .310 and was a good fielder and very fast runner. Hafey had a lifetime batting average of .317 and a slugging average of .526. He was an exceptionally good outfielder with speed and a strong arm. Lou Brock—a good hitter but poor fielder—was probably elected based mainly on his terrific speed. He broke Ty Cobb's all-time stolen base record and today ranks second all-time. Speed was also an important factor in the election of Ricky Henderson, the all-time leader in stolen bases. Thus, of the 21 left fielders elected to the Hall of Fame, 17 were probably elected based on their hitting, 2 based on their hitting and fielding, 1 based on a combination of his hitting and speed, and 1 based on his speed alone.

Right Fielders

Right fielders tend to be more like left fielders than center fielders. Fielding is not as important versus hitting for right fielders and left fielders as it is for center fielders. As a result, right fielders and left fielders tend to be slower, bigger, and better hitters than center fielders. Right fielders rank in the middle in size and speed and near the top in hitting.

In Table 9.1, right fielders rank from third to fifth in size and from third to sixth in speed, depending on the period of time. From 1876 to 1941 right fielders stole a base in 13% of their games, from 1942 to 2006 in 7% of their games, and in 2007 in 5% of their games. From 1876 to 1941 right fielders and center fielders averaged 5 feet 10 inches and 174 pounds (compared with 5 feet 11 inches and 180 pounds for left fielders), from 1942 to 2006 right fielders averaged 6 feet 1 inch and 192 pounds (vs. 6 feet 1 inch and 201 pounds for left fielders and 6 feet and 188 pounds for center fielders), and in 2007 they averaged 6 feet 2 inches and 212 pounds (vs. 6 feet 1 inch and 209 pounds for left fielders and 6 feet and 196 pounds for center fielders). In table 9.2, the number of right fielders among the top 50 hitters of all time ranks from first to third in eight of the nine models. For only the RC/27 model is the number of right fielders as low as fourth.

The top 10 hitting right fielders of all time according to the different measures of hitting performance are identified in table 9.17. If you look at the top 10 right fielders as a group, there is general agreement on only five who belong there. All of the measures agree on two (Ruth and Walker), all except one measure agree on another (Ott), and all except two measures agree on two more (Guerrero and Heilmann). Twenty-one right fielders share the top 10 rankings, and 14 of them share the top five rankings.

If you look at the specific rankings of the right fielders within the top 10, there is more disagreement than agreement. All measures agree that Ruth is number 1, and four measures agree that Walker is number 2. None of the 19 remaining right fielders have more than two measures that agree on their specific ranking.

TABLE 9.17 The Top 10 Hitting Right Fielders of All Time According to Various Measures

Player	PRG	Advanced Weighted Measures			Basic Weighted Measures			Unweighted Measures	
		LSLR	RC/27	LWTS	TA	OPS	SLG	OBP	AVG
Babe Ruth	1	1	1	1	1	1	1	1	1
Sam Thompson	2	2	3	5	—	—	—	—	6
Harry Heilmann	3	—	6	10	5	5	—	3	2
Larry Walker	4	7	2	2	2	2	3	8	—
Juan Gonzalez	5	—	—	—	—	—	4	—	—
Mel Ott	6	8	5	3	3	4	9	2	—
Vladimir Guerrero	7	3	—	4	7	3	2	—	8
Chuck Klein	8	—	7	—	—	8	6	—	10
Babe Herman	9	—	—	—	—	9	10	—	7
Hank Aaron	10	5	10	9	—	6	5	—	—
Frank Robinson	(11)	9	8	8	6	7	7	—	—
Gary Sheffield	(13)	6	—	—	9	10	—	9	—
Bobby Abreu	(14)	10	9	7	4	—	—	6	—
Paul Waner	(29)	—	—	—	—	—	—	5	4
Mike Tiernan	(32)	4	4	6	8	—	—	10	—
Other players[1]	—	—	—	—	—	—	—	—	—

[1] Six other right fielders were ranked in the top 10 but in one measure only: J. D. Drew #10 in TA, Sammy Sosa #8 in SLG, Roy Cullenbine #4 and Brian Giles #7 in OBP, and Tony Gwinn #3, Ichiro Suzuki #5, and Kiki Cuyler #9 in AVG.

Table 9.18 summarizes the normalized PRG ratings for the top 10 hitting right fielders. One right fielder who played most of his career in the Era of Constant Change (Sam Thompson) and another right fielder who played in the Live Ball Interval (Mike Tiernan) are excluded from the table for reasons given previously.

The biggest gainer by far from the normalization process was Sam Crawford, who gained 27 positions. Crawford played during the Dead Ball Era when pitchers had the upper hand and PRG averages were at their lowest. His batting accomplishments, therefore, were much greater than his raw statistics indicate. Jack Clark gained 15 positions. He played during the Live Ball Revived Era but spent most of his time in the National League, where PRG averages were

TABLE 9.18 Normalizing the PRG Ratings of the Top 10 Hitting Right Fielders

Before Normalization			Normalization		After Normalization		
			LG				
Rank	Rating	Player	PRG	NORM[1]	Rank	Rating	Player
1	1.480	Babe Ruth	.933	1.586	1	1.586	Babe Ruth*
2	1.233	Harry Heilmann	.915	1.348	2	1.356	Hank Aaron*
3	1.230	Larry Walker	.929	1.324	3	1.354	Frank Robinson*
4	1.226	Juan Gonzalez	.974	1.259	4	1.348	Harry Heilmann*
5	1.223	Mel Ott	.908	1.347	5	1.347	Mel Ott*
6	1.219	Vladimir Guerrero	.979	1.246	6	1.324	Larry Walker
7	1.194	Chuck Klein	.923	1.294	7	1.318	Sam Crawford*[2]
8	1.191	Babe Herman	.931	1.279	8	1.294	Chuck Klein*
9	1.178	Hank Aaron	.869	1.356	9	1.279	Babe Herman
10	1.171	Frank Robinson	.865	1.354	10	1.269	Jack Clark[3]

* Indicates that a player was elected to the Baseball Hall of Fame.
[1] Player's PRG rating divided by his league's PRG rating for the years in which he played.
[2] Before normalization Crawford was ranked 29th with a PRG rating of 1.085 vs. a league average of .823.
[3] Before normalization Clark was ranked 25th with a PRG rating of 1.109 vs. a league average of .874.

lower than in the American League. Hank Aaron and Frank Robinson gained seven positions each. They played during the Dead Ball Interval, another period when pitchers dominated batters and PRG averages were low.

The biggest losers from normalization were Vladimir Guerrero and Juan Gonzalez, dropping nine and eight positions, respectively, because they played when league PRG averages were high. Larry Walker lost three positions, Harry Heilmann lost two positions, and Chuck Klein and Babe Herman lost one position each. The rankings of Babe Ruth and Mel Ott remain unchanged.

Two other active right fielders have a chance to break into the top 10. Gary Sheffield is currently number 14, and Vladimir Guerrero is number 15. Guerrero has the best chance to advance because he is only 35 years old and Sheffield is over 40.

More right fielders have been elected to the Baseball Hall of Fame than players at any other position except pitching—a total of 24, including the seven identified in table 9.18. The election of 11 of the other 17 must also have been based primarily, if not exclusively, on their hitting. Nine of those 11—Sam Thompson, Willie Keeler, Tony Gwynn, Paul Waner, Sam Rice, Ross Youngs, Kiki Cuyler, Roberto Clemente, and Elmer Flick—had lifetime batting averages ranging from .313 to .341 and ranked from third to 14th all-time for right fielders. Two others—Reggie Jackson and Dave Winfield—were also undoubtedly elected for their slugging prowess. Only six right fielders were elected to the Hall of Fame based on something other than their hitting. King Kelly was elected based on his all-around ability. Tommy McCarthy and Harry Hooper were outstanding defensive outfielders with mediocre hitting credentials. Enos Slaughter's election was probably based on his great base running because his hitting and fielding were good, but not great. Al Kaline and Andre Dawson were good players, both at the plate and in the field. Thus, of the 24 right fielders elected to the Hall of Fame, 18 were probably elected based mainly on their hitting, 2 were probably elected based primarily on their fielding, 3 were probably elected based on a combination of factors, and 1 was probably elected based mainly on his base running.

Designated Hitters

Designated hitters were not included in table 9.1 because the designated hitter rule was not adopted by the American League until 1973. For the 37 years that the designated hitter rule has been in effect, designated hitters tend to be among the biggest and slowest players in American League lineups. The 15 designated hitters in table 9.19 were about average height (6 feet 1 inch) but, along with first basemen and catchers, were the heaviest players (206 pounds). They were faster than third basemen, first basemen, and catchers but slower than all of the other position players. Their average speed was bolstered because three of them had 200 or more stolen bases in their careers—Paul Molitor (504), Don Baylor (285), and Jose Canseco (200). In 2009, 16 designated hitters played in 50 or more games— they averaged 6 feet 2 inches tall and 226 pounds. Heightwise, they were exceeded only by first basemen, and they were the heaviest position players in American League lineups. They were slower than

TABLE 9.19 The Top 10 Designated Hitters of All-Time According to Various Measures

Player	PRG	Advanced Weighted Measures			Basic Weighted Measures			Unweighted Measures	
		LSLR	RC/27	LWTS	TA	OPS	SLG	OBP	AVG
Frank Thomas	1	1	1	1	1	1	1	1	3
David Ortiz	2	4	4	3	3	3	2	3	7
Edgar Martinez	3	2	2	2	2	2	4	2	1
Jose Canseco	4	3	7	4	4	4	3	—	—
Mike Sweeney	5	7	5	7	5	5	5	6	4
Cliff Johnson	6	—	9	—	8	8	8	—	—
Harold Baines	7	10	—	10	—	6	7	9	6
Andre Thornton	8	9	10	8	7	10	—	7	—
Hal McRae	9	10	—	5	—	—	9	—	5
Chili Davis	10	6	—	9	10	—	—	8	9
Aubrey Huff	(11)	6	3	—	—	9	6	—	8
Oscar Gamble	(12)	—	8	—	9	—	10	10	—
Don Baylor	(13)	8	—	—	—	—	—	—	—
Brian Downing	(14)	—	—	—	—	—	—	4	10
Paul Molitor	(15)	5	6	6	6	7	—	5	2

all except catchers and first basemen. In 2009, only 7 of the 16 who played in 50 or more games had any stolen bases.

The data also show that designated hitting on most teams tends to be shared by many players. In 2009, a total of 162 American League players were designated hitters in at least one game—for an average of 11-plus designated hitters per team. A total of 50 National League players were designated hitters in interleague games—for an average of three-plus designated hitters per team.

The top 10 hitting designated hitters for the years in which the designated hitter rule has been in effect in the American League are identified in table 9.19. If you look at the top 10 designated hitters as a group, there is general agreement on nine who belong in the top 10 (the most for any position). All of the measures agree on four of them (Thomas, Ortiz, Martinez, and Sweeney), and all except two measures agree on four more (Canseco, Baines, Thornton, and Molitor). Fifteen designated hitters share the top 10 rankings, and nine of them share the top five rankings.

If you look at the specific rankings of the designated hitters within the top 10, there is some disagreement. All except one measure agree that Thomas is number 1, six measures agree that Martinez is number 2, five measures agree that Sweeney is number 5, and four measures agree that Ortiz should be number 3 and Canseco should be number 4. None of the remaining 10 designated hitters have more than three measures that agree on their specific rankings.

The designated hitter rankings in table 9.19 are based on raw data. Table 9.20 normalizes the PRG ratings. Normalization has less of an impact on designated hitters because the designated hitter rule has been in effect for only the last 37 years. Thus, the rankings of the top 4 of the 10 players in table 9.20 do not change at all. The ranking of three players changes five places—Harold Baines and Mike Sweeney lose and Don Baylor gains. Andre Thornton and Hal McRae gain two places each, and Cliff Johnson and Chili Davis lose one place each.

TABLE 9.20 Normalizing the PRG Ratings of the Top 10 Designated Hitters

Before Normalization			Normalization		After Normalization		
Rank	Rating	Player	LG PRG	NORM[1]	Rank	Rating	Player
1	1.268	Frank Thomas	.975	1.301	1	1.301	Frank Thomas
2	1.234	David Ortiz	.988	1.249	2	1.249	David Ortiz
3	1.193	Edgar Martinez	.971	1.229	3	1.229	Edgar Martinez
4	1.149	Jose Canseco	.958	1.199	4	1.199	Jose Canseco
5	1.134	Mike Sweeney	.985	1.151	5	1.198	Cliff Johnson
6	1.091	Cliff Johnson	.911	1.198	6	1.179	Andre Thornton
7	1.084	Harold Baines	.948	1.143	7	1.177	Hal McRae
8	1.074	Andre Thornton	.911	1.179	8	1.157	Don Baylor[2]
9	1.064	Hal McRae	.904	1.177	9	1.155	Oscar Gamble[3]
10	1.058	Chili Davis	.925	1.144	10	1.151	Mike Sweeney

[1] Player's PRG rating divided by his league's PRG rating for the years in which he played.

[2] Before normalization Baylor was ranked 13th with a PRG rating of 1.023 vs. a league average of .884.

[3] Before normalization Gamble was ranked 12th with a PRG rating of 1.035 vs. a league average of .896.

TABLE 9.21 Estimated Basis for Election to the Baseball Hall of Fame

Position	Hitting No.	Hitting (Pct.)	Other No.	Other (Pct.)	Total
Designated hitters	1	100	—	—	1
First basemen	17	94	1	6	18
Left fielders	16	80	4	20	20
Right fielders	18	75	6	25	24
SUBTOTAL	52	83	11	17	63
Center fielders	11	65	6	35	17
Third basemen	6	60	4	40	10
Catchers	7	54	6	46	13
SUBTOTAL	24	60	16	40	40
Second basemen	9	50	9	50	18
Shortstops	8	36	14	64	22
ALL PLAYERS	93	65	50	35	143

Position Players in the Hall of Fame

Each of the preceding player position sections has speculated on the primary reason for election of players to the Hall of Fame (hitting, fielding, etc.)—speculated because the baseball literature contains no official tabulation. Table 9.21 summarizes and compares the results by position. The data in this table parallel those in table 9.2, which listed the top 50 hitters of all time by position. The positions with the most top 50 hitters tend to be the same ones in which hitting is the predominant factor most often in elections to the Hall of Fame. Hitting was the predominant factor in the election of 94% of first basemen to the Hall of Fame, the most for any position player. Among the outfielders, hitting was the predominant factor for 75% of right fielders and 80% of left fielders, but only 65% of center fielders. The positions with the fewest top 50 hitters tend to be the same ones in which hitting is the predominant factor least often. Hitting was the predominant factor in the election of 50% of second basemen and 36% of shortstops. It is interesting that the so-called skilled fielding positions—those who play up the middle on defense, that is, catcher, shortstop, second baseman, and center fielder—are the ones for which hitting is least often the predominant factor in elections to the Hall of Fame.

* 10 *

The Table Setters

Once begun, a task is easy; half the work is done.

Horace, *Epistles* (20–c. 8 BC)

A good beginning makes a good ending.

English proverb

....

The Leadoff Batters

The expectations and opportunities for leadoff batters are different from those for players in the middle of the batting order.[1] Leadoff batters are expected to get on base frequently and, once there, to run the bases well. Stealing bases is only one of the criteria for running the bases well. Joe McCarthy said Joe DiMaggio was the greatest base runner he ever saw, yet DiMaggio stole only 30 bases in his entire major league career! A more important criterion for running the bases well may be advancing an extra base on balls hit to the outfield. In the very early days of baseball (from 1886 to 1898) the statistics for stolen bases included advancing an extra base on balls hit to the outfield. Caught stealing is a very important negative criterion in base running—doubly so because not only is the base runner removed from the bases but his teammates have one less out to use for the remainder of an inning. Leadoff power is desirable, but not necessary, because leadoff batters have fewer opportunities to advance runners and to drive in runs. Only 6 of the 88 leadoff batters in this study had more than 200 career home runs, and only 2 of them had a home run average (HR ÷ AB) of more than 3%.[2]

Switch-hitting is more common among leadoff batters than middle-of-the-order batters. Taking advantage of the percentages favoring right-handed batters against left-handed pitchers and left-handed batters against right-handed pitchers is apparently easier for batters just trying to get on base than for batters trying to drive the ball a long way. Twenty-three of the 88 leadoff batters (26%) were switch-hitters. The remaining 65 leadoff batters were about evenly divided between left- and right-handed batters.

Leadoff batters tend to be smaller and quicker or faster than middle-of-the-order batters and most often come from the ranks of positions that place a premium on these qualities in the field—second base, shortstop, and center field. Fifty-four of the 88 leadoff batters (61%) played one of those three positions in the field. Very few catchers, right fielders, or designated hitters are leadoff batters. None of the leadoff batters in this study were catchers, only one was a designated hitter, and there were only three right fielders. Leadoff batters have always been smaller than other position players—1 or 2 inches shorter and 10–30 pounds lighter, depending on the historical era. The difference in size has become more pronounced in the two most recent eras.

Many in the baseball community do not fully appreciate the accomplishments of leadoff batters. Only 16 of the more than 200 position players elected to the Baseball Hall of Fame were leadoff batters. One reason for this lack of appreciation is the way batting statistics are compiled and presented. The statistics for leadoff batters are lumped together with the statistics for all other batters instead of being presented separately. The popular statistics purportedly related to leadoff batter performance—batting average, on-base percentage, stolen bases, and caught stealing—should be included in separate statistical summaries for leadoff batters. Most importantly, these listings should also include PRG and TBBA (total bases plus walks and hit by pitch divided by plate appearances).

Table 10.1 summarizes the rankings of the top 20 leadoff batters of all time according to different measures. Two measures that were not included in the previous tables on middle-of-the-order batters have been added: total batter bases average (TBBA) and expected runs scored (ERS), Bill James's formula constructed especially for leadoff batters.[3] Billy Hamilton ranks first or second and John McGraw ranks first, second, or third in 9 of the 11 measures; yet they rank lowest in ERS. Bill James "knocked down" his ERS rankings of Billy Hamilton and John McGraw to #19 and #20, respectively, because they played in an era when so many runs were scored. A major reason why so many runs were scored was the high rate of errors at that time. More errors meant more runners on base for batters to advance or drive in.

If you look at the top 20 leadoff batters as a group, there is general agreement on 10 who belong in the top 20. All of the measures agree on three (Hamilton, Boggs, and Raines), all except one measure agree on four (McGraw, Burkett, Combs, and Henderson), and all except two measures agree on three more (Galan, Molitor, and Ryan). Thereafter, however, the agreement quickly degenerates. Of the remaining 36 leadoff batters, only 5—Bobby Bonds, Carew, Blue, Dom DiMaggio, and Lofton—are ranked in the top 20 by seven or more of the measures.

If you look at the specific ranking of the leadoff batters within the top 20, there is more disagreement. Seven measures agree that Hamilton is number 1, five measures agree that McGraw is number 2, and four measures agree that Burkett is number 3 and Boggs is number 5. Only 1 of the remaining 42 leadoff batters—Bonds—has as many as three measures that agree on his specific ranking. Twenty-eight of the 46 leadoff batters do not have any two measures that agree on their specific ranking.

There is, however, one qualification that needs to be made with regard to the use of the PRG measure to rate leadoff batters. The three-step process in scoring runs—getting on base, advancing runners, and driving in runs—is the same for leadoff batters as it is for middle-of-the-order batters. There is a difference between the two groups of players, but it pertains not to the process itself but to the

TABLE 10.1 The Top 20 Leadoff Batters of All Time According to Various Measures

Player	PRG	Advanced Weighted Measures				Basic Weighted Measures				Unweighted Measures	
		LSLR	RC/27	ERS	LWTS	TBBA	TA	OPS	SLG	OBP	AVG
Billy Hamilton	1	1	1	19	1	2	1	1	10	2	1
Wade Boggs	2	12	11	5	5	6	11	5	6	4	5
John McGraw	3	2	2	20	2	3	2	2	—	1	3
Augie Galan	4	—	12	7	15	8	16	13	18	15	—
Bobby Bonds	5	7	—	6	12	1	6	6	1	—	—
Earle Combs	6	6	9	—	8	5	8	4	2	10	7
Paul Molitor	7	9	19	—	11	12	18	10	3	—	12
Jimmy Ryan	8	7	4	—	6	10	7	9	5	—	10
Jesse Burkett	9	3	3	—	3	7	4	3	4	4	2
Lou Whitaker	10	—	—	—	—	15	—	—	12	—	—
Rod Carew	11	19	—	13	—	—	—	7	11	14	6
Johnny Damon	12	14	—	—	—	17	—	18	7	—	—
Lu Blue	13	16	—	—	17	18	12	14	—	8	—
Rickey Henderson	14	5	10	1	4	4	3	8	20	9	—
Tim Raines	15	17	13	2	7	9	5	12	14	17	20
Elbie Fletcher	16	—	—	15	—	—	—	—	—	18	—
Brady Anderson	17	—	—	—	—	10	15	—	13	—	—
Dom DiMaggio	18	11	—	—	—	19	—	15	19	20	17
Kenny Lofton	19	13	—	—	13	—	14	17	15	—	16
Lonnie Smith	20	—	—	17	—	20	—	19	16	—	—
Craig Biggio	(21)	18	—	8	—	13	—	16	9	—	—
Tommy McCarthy	(22)	15	6	—	14	—	20	—	—	—	—
Ichiro Suzuki	(23)	4	14	—	10	—	—	11	8	—	4
Max Bishop	(27)	—	—	—	20	14	10	—	—	3	—
Pete Rose	(28)	—	—	10	—	—	—	—	—	—	13
Stan Hack	(30)	—	18	14	16	—	—	20	—	12	14
Len Dykstra	(33)	—	—	4	19	16	13	18	—	—	—
Dummy Hoy	(37)	10	8	—	9	—	9	—	—	16	—
Eddie Stanky	(37)	—	—	9	—	—	—	—	—	7	—
Paul Hines	(39)	—	5	—	—	—	—	—	—	—	15
Richie Ashburn	(41)	—	—	—	—	—	—	—	—	11	9
Topsy Hartsel	(54)	—	20	3	—	—	19	—	—	18	—
Roy Thomas	(67)	—	15	12	18	—	17	—	—	6	—
Other players[1]	—	—	—	—	—	—	—	—	—	—	—

[1] Thirteen other players are ranked in the top 20 but in one measure only: Ron LeFlore #20 in LSLR; Abner Dalrymple #7, Arlie Latham #15, and Tom Brown #17 in RC/27; Don Buford #11, Miller Huggins #16, and Bob Bescher #18 in ERS; Juan Samuel #17 in SLG; Eddie Yost #12 in OBP; and Lloyd Waner #8, Matty Alou #11, Mickey Rivers #18, and Bip Roberts #19 in AVG.

opportunities to participate in that process. The opportunities to participate are not equal—leadoff batters have many fewer opportunities to advance runners and drive in runs. The solution is not to abandon PRG as a measure for leadoff batters, but to apply a different PRG scale for judging the performance of leadoff batters. While a PRG rating of 1.020 would not be very high for a middle-of-the-order batter, it would be very high for a leadoff batter. Do not compare the PRG ratings of leadoff batters with the PRG ratings of middle-of-the-order batters. Treat both groups separately.

Table 10.2 summarizes the normalized PRG ratings for the top 20 leadoff batters of all time. Leadoff batters who played most or all of their careers in the Era of Constant Change and/or Live Ball Interval are excluded from this table. The fact that the very rules of the game itself were so different in the first two eras, particularly in the Era of Constant Change, made it impossibly difficult to compare the performance of players then with that of those who played later. Excluding those players from table 10.2 does not mean that they were not as good as those in subsequent eras. It simply means that a reasonable comparison of the statistics is not possible. Billy Hamilton, John McGraw, Jimmy Ryan, and Jesse Burkett were certainly excellent leadoff batters and may have been better than many, if not most, of the other leadoff batters in table 10.2, but that is something we do not know now and may never know with any certainty.

The biggest gainers by far from normalization were Topsy Hartsel and George Burns, who played during the Dead Ball Era when the balance favored pitchers over batters. Their batting accomplishments were actually greater than their raw numbers would indicate. Pete Rose also gained significantly in the normalization process because most of his career was spent in the Dead Ball Interval. The biggest losers from normalization were Johnny Damon, Brady Anderson, Kenny Lofton, and Ichiro Suzuki. They played during the Live Ball Enhanced Era when the balance favored batters over pitchers. Their batting accomplishments were thus less than their raw numbers would indicate.

Since there are no active players in the top 20 list, the all-time rankings for leadoff batters are likely to remain the same for the fore-

TABLE 10.2 Normalizing the PRG Ratings of the Top 20 Hitting Leadoff Batters

	Before Normalization		Normalization			After Normalization	
			LG				
Rank	Rating	Player	PRG	NORM[1]	Rank	Rating	Player
1	.995	Wade Boggs	.945	1.053	1	1.116	Bobby Bonds
2	.985	Augie Galan	.894	1.102	2	1.102	Augie Galan
3	.980	Bobby Bonds	.878	1.116	3	1.068	Rod Carew*
4	.971	Earle Combs	.963	1.008	4	1.054	Elbie Fletcher
5	.963	Paul Molitor	.936	1.029	5	1.053	Wade Boggs*
6	.944	Lou Whitaker	.930	1.015	6	1.029	Paul Molitor*
7	.943	Rod Carew	.883	1.068	7	1.029	Tim Raines
8	.939	Johnny Damon	.985	.953	8	1.018	Lonnie Smith
9	.939	Lu Blue	.956	.982	9	1.017	Pete Rose[2]
10	.936	Rickey Henderson	.939	.997	10	1.015	Lou Whitaker
11	.934	Tim Raines	.908	1.029	11	1.011	Topsy Hartsel[3]
12	.931	Elbie Fletcher	.883	1.054	12	1.008	Earle Combs*
13	.918	Brady Anderson	.968	.948	13	.997	Rickey Henderson*
14	.916	Dom DiMaggio	.920	.996	14	.996	Dom DiMaggio
15	.904	Tony Phillips	.948	.954	15	.982	Lu Blue
16	.899	Lonnie Smith	.883	1.018	16	.979	Stan Hack[4]
17	.899	Kenny Lofton	.971	.926	17	.976	George Burns[5]
18	.894	Craig Biggio	.919	.973	18	.973	Craig Biggio
19	.886	Ichiro Suzuki	.974	.910	19	.970	Juan Samuel[6]
20	.883	Tony Fernandez	.934	.945	20	.956	Pee Wee Reese*[7]

* Indicates that a player was elected to the Baseball Hall of Fame.

[1] Player's PRG rating divided by his league's PRG rating for the years in which he played.

[2] Before normalization Rose was ranked 23rd with a PRG rating of .874 vs. a league average of .859.

[3] Before normalization Hartsel was ranked 37th with a PRG rating of .816 vs. a league average of .807.

[4] Before normalization Hack was ranked 25th with a PRG rating of .871 vs. a league average of .890.

[5] Before normalization Burns was ranked 31st with a PRG rating of .828 vs. a league average of .848.

[6] Before normalization Samuel was ranked 26th with a PRG rating of .866 vs. a league average of .893.

[7] Before normalization Reese was ranked 28th with a PRG rating of .859 vs. a league average of .899.

seeable future. The highest-ranked leadoff batter currently active is Johnny Damon (#22). He is 37 years old and has some time left to improve his numbers enough to break into the top 20.

Only 6 of the top 20 players in table 10.2 have been elected to the Baseball Hall of Fame: Combs, Boggs, Molitor, Carew, Henderson, and Reese. Craig Biggio retired at the end of the 2007 season. Thus,

13 of the 19 eligible players in table 10.2 have been passed over in the voting for the Hall of Fame. Tim Raines has received enough votes to remain on the ballot and could be elected in the near future, as could Biggio when he becomes eligible.

Four of the six Hall of Fame leadoff batters in table 10.2—Rod Carew, Wade Boggs, Paul Molitor, and Pee Wee Reese—were probably elected based on their hitting. Ricky Henderson was probably elected based on his hitting and speed, and Earle Combs based on his hitting and fielding. Ten leadoff batters not in table 10.2 were also elected to the Hall of Fame. Jesse Burkett and Billy Hamilton were undoubtedly elected based on their hitting because they were great hitters and not known for their fielding prowess. Second baseman Red Schoendienst and shortstops Luis Aparacio and Phil Rizzuto were undoubtedly elected based more on their defensive skills than their batting skills. Dave Bancroft had higher batting statistics and was probably elected based more on his hitting. Center fielders Richie Ashburn and Lloyd Waner were probably elected based on their all-around abilities since they were good hitters, fielders, and base runners. Center fielder Max Carey's election was undoubtedly based on his fielding alone because he was one of the best outfielders of his day and a weak hitter. Tommy McCarthy is one of only three right-field leadoff batters in this study—the other two being Bobby Bonds and Ichiro Suzuki.[4] McCarthy was an excellent defensive right fielder and base runner. Two other leadoff batters—John McGraw and Miller Huggins—were elected to the Hall of Fame, but as managers rather than players. Huggins is ranked in the top 20 in only one of the measures (ERS) summarized in table 10.1. Thus, seven leadoff batters were probably elected to the Baseball Hall of Fame based on their hitting, two based on their hitting and fielding or hitting and speed, four based on their fielding, one based on his fielding and speed, and two based on their all-around performance (hitting, fielding, and speed).

One of the most exclusive "clubs" in the baseball world is the 3,000-hit club. Only 26 players in the entire history of Major League Baseball are members of that club. While getting on base frequently is a major objective of leadoff batting, I have not included the number of hits as a measure in table 10.1 because it is a quantitative measure tied closely to longevity rather than a qualitative measure tied to player opportunities. Thus, it is interesting to note that only 6 of the 46 players in table 10.1 have 3,000 or more hits—Rose (4,256), Molitor (3,319), Biggio (3,060), Henderson (3,055), Carew (3,053), and Boggs (3,010). While 3,000 or more hits is certainly a great accomplishment, it is a quantitative measure based partly on player endurance. If we measure their 3,000-plus hits from a qualitative perspective, most of their batting averages are certainly good—Boggs (.328), Carew (.328), Molitor (.306), Rose (.303), Biggio (.281), and Henderson (.279)—but only Boggs and Carew rank in the top 100 all-time (29th and 30th, respectively). The fact that all six of these players ranked in the top 20 in PRG after normalization is due less to the number of hits they had than to the number of bases they produced.

Before concluding this section, we need to recognize the special accomplishments of the leading leadoff batters of all time. Bobby Bonds played from 1968 to 1981: seven years with the San Francisco Giants, followed by seven years split between seven different teams. He was a new kind of leadoff batter—a big (6 feet 1 inch and 190 pounds) right-handed-hitting right fielder who had speed on the bases to go with his power at the plate. Bonds was the first player in major league history with 30 or more home runs and 30 or more stolen bases in a season, and he repeated this feat four times for a major league record that still stands. He is the only leadoff batter in the history of Major League Baseball with more than 300 home runs (332) and more than 400 stolen bases (461). The only middle-of-the-order batter with over 300 home runs and 400 stolen bases is Bobby's son Barry Bonds, who has 762 home runs and 514 stolen bases. The combination of his power and speed made Bobby Bonds the greatest leadoff batter of all time. The fact that Bobby Bonds has never been elected to the Hall of Fame is the prime example of a leadoff batter's accomplishments being underappreciated.

Augie Galan played from 1934 to 1949: eight years with the Chicago Cubs, followed by five years with the Brooklyn Dodgers and three years split between three other teams. Galan was a switch-hitting left fielder. He did not have spectacular numbers, but he played well throughout his career. In fact, he actually played better in his 30s than in his 20s. He had an OBP of over .400 six years in a row, for a combined OBP of .439, an exceptionally good performance for any player. Galan holds two distinctive firsts: the first regular major league player to complete a season without hitting into a double play, and the first National League player to hit a left-handed and a right-handed home run in the same game. The fact that Augie Galan has never been elected to the Hall of Fame is another example of a leadoff batter whose accomplishments have not been sufficiently appreciated.

Rod Carew played from 1967 to 1985: 12 years with the Minnesota Twins, followed by seven years with the California Angels. Carew was a left-handed leadoff batter who played second base the first nine years of his major league career. Because he was not a great fielder, he was then transferred to first base, which he played the rest of his career. Carew was a great place hitter, bunter, and base runner. He batted over .300 15 consecutive seasons (for a combined batting average of .337), an achievement exceeded by only Tony Gwynn and equaled by only five other players in the entire history of Major League Baseball. Among Rod Carew's achievements were Rookie of the Year, one Most Valuable Player Award, 18 straight All-Star Game selections, seven batting titles, and a first-time selection to the Hall of Fame in 1991—the highest-ranked leadoff batter (after normalization) in the Hall of Fame.

Elbie Fletcher played from 1934 to 1949: five-plus seasons with the Boston Braves, followed by six-plus seasons with the Pittsburgh Pirates, and finally one last season back with the Braves. Fletcher was an average-sized (6 feet and 180 pounds) left-handed-batting first baseman. He did not have power nor was he a fast base runner, but he did have the ability to get on base often. His best seasons were with the Pirates. In one five-year stretch, Fletcher averaged 110 walks per season and led the league in OBP three seasons in a row. He has never been elected to the Hall of Fame.

Wade Boggs played from 1982 to 1999: his first 11 years were with the Boston Red Sox, followed by five years with the New York

Yankees and two years with the Tampa Bay Devil Rays. Boggs was a good-sized (6 feet 2 inches and 197 pounds) left-handed-hitting third baseman. He had the highest batting average for a rookie (.349) in the history of the American League. Boggs was not a fast runner nor was he a power hitter, but he did have consistently high batting averages and OBPs. He batted over .300 each of his first 10 years and again in four of his last eight years. Boggs had an OBP of over .400 his first eight years and again in three subsequent years. He led the league in AVG five times and in OBP six times. Wade Boggs was a first-time selection for the Hall of Fame in 2005—the second-highest-ranked leadoff batter (after normalization) in the Hall of Fame.

Paul Molitor played from 1978 to 1998: 15 years with the Milwaukee Brewers, followed by three years with the Toronto Blue Jays and three years with the Minnesota Twins. Molitor was a right-handed line-drive hitter. He was unique (a designated hitter who batted leadoff) and versatile (a designated hitter who could play several positions in the field). Early in his career he played third base and second base but was switched to designated hitter to keep him off the disabled list and in the batting order. Molitor did not have a powerful bat (234 career home runs and a lifetime slugging average of only .448), but it was a steady bat. He batted over .300 12 times, for a lifetime average of .306, and had an on-base percentage of .350 or more 14 times, for a career mark of .369. Molitor led the league in runs and hits three times each and in doubles and triples once each, but he never led the league in any average category. He had some speed—he averaged 24 stolen bases per year, for a total of 504. Molitor was selected to play in six All-Star Games and was elected to the Hall of Fame in 2004, his first year of eligibility. Paul Molitor is the only designated hitter and the third-highest-ranked leadoff batter (after normalization) elected to the Hall of Fame.

Tim Raines played from 1979 to 2002: 12 years with the Montreal Expos, five years with the Chicago White Sox, three years with the New York Yankees, and at the end of his career three years with four other teams. Raines was a small (5 feet 8 inches and 178 pounds) but well-built—he was nicknamed "Rock"—switch-hitting left fielder. Speed was his forte. He stole 808 bases and was caught stealing only

146 times, for a stolen base percentage of 84.7, the third highest ever. Carlos Beltran and Pokey Reese had higher percentages but combined had less than half the attempted steals of Raines. In his first 12 seasons as a regular player, he averaged 60 stolen bases and had 70 or more stolen bases six years in a row. Raines led the league in stolen bases four years in a row, but he seldom led the league in any other event—twice in runs scored and once each in doubles, batting average, and on-base percentage. He was selected to play in seven All-Star Games. In January 2008, Tim Raines was the only first-time candidate for election to the Hall of Fame to receive enough votes to remain on the ballot for future consideration, and he remained on the ballot after the 2009 selections.

Lonnie Smith played from 1978 to 1994: four seasons with the Philadelphia Phillies, followed by three seasons with the St. Louis Cardinals, three seasons with the Kansas City Royals, five seasons with the Atlanta Braves, one season with the Pittsburgh Pirates, and one season with the Baltimore Orioles. Smith was a right-handed-batting left fielder who did not play regularly for much of his career—he had only one season with more than 600 plate appearances—but he had his moments. In one five-year stretch he averaged 125 hits, 39 walks, 43 stolen bases, and 78 runs. These are not great numbers, but they were achieved with an average of only 412 at bats. His best year was 1989, when he had a batting average of .315, on-base percentage of .415, and slugging average of .533. He led the league in runs and on-base percentage once each. Smith was selected to play in one All Star Game and has never been elected to the Baseball Hall of Fame.

Pete Rose had the longest career of any player ever if you measure careers by the number of games played, plate appearances, or at bats. He played 24 years from 1963 to 1986: 16 years with the Cincinnati Reds, five years with the Philadelphia Phillies, one year with the Montreal Expos, and two years back with Cincinnati. Rose was an average-sized (5 feet 11 inches and 200 pounds), aggressive (nicknamed "Charlie Hustle") switch-hitting batter and versatile fielder who played over 600 games at each of four different positions (first base, second base, third base, and left field) and nearly 600 games at a fifth position (right field). He led the league in hits seven times, doubles five times, runs four times, batting average three times, and on-base percentage twice. Rose holds the all-time record for hits and ranks second in doubles and sixth in runs. He was named Rookie of the Year, holds one Most Valuable Player Award, and was selected to play in 17 All-Star Games.

Rose's impressive career on the field was eclipsed by his off-the-field activities. He was convicted of income tax evasion and served five months in a federal prison. Pete Rose was banned from organized baseball and declared ineligible for election to the Hall of Fame because he bet on baseball games.

Lou Whitaker played from 1977 to 1995, spending all 19 of his seasons with the Detroit Tigers. For most of that time, he and Alan Trammell teamed together both in the field and at the plate. Whitaker was the second baseman and Trammell the shortstop. Whitaker was the leadoff batter and Trammell batted second. They were excellent fielders, but neither ever led the American League in any major batting event. Whitaker was a slugging leadoff batter, with a slugging average over .450 seven times and 244 career home runs. He was Rookie of the Year, won three Gold Glove Awards, and was selected to play in the All-Star Game five times. Lou Whitaker has not been elected to the Baseball Hall of Fame.

Topsy Hartsel played from 1898 to 1911: three part-time seasons (two with the Louisville Colonels and one with the Cincinnati Reds), two full-time seasons (one with Cincinnati and another with the Chicago Cubs), followed by 10 seasons with the Philadelphia Athletics. Hartsel was a small (5 feet 5 inches and 156 pounds) left-handed-hitting left fielder. He was not a slugger—only 31 home runs in his entire 14-year career. Hartsel's forte was getting on base, finishing with a career .384 on-base percentage. His career on-base percentage was actually greater than his career slugging average. Hartsel's first eight years with Philadelphia were his prime years—he averaged 141 hits, 89 runs, 87 walks, 28 stolen bases, and 20 doubles each year. He led the American League in walks five times, on-base percentage twice, and runs and stolen bases once each. Hartsel has never been elected to the Baseball Hall of Fame.

Earle Combs played from 1924 to 1935, spending his entire 12-

year career with the New York Yankees. He was the leadoff batter and center fielder for the Yankees during the Babe Ruth / Lou Gehrig era. Combs was a left-handed line-drive hitter who also walked frequently. He had a lifetime batting average of .325 and an on-base percentage of .397. Combs was a fast runner—he led the American League in triples three times—but wasn't called upon to steal very often with the likes of Ruth and Gehrig behind him in the batting order. In one eight-year stretch, he averaged 125 runs per season. In 1927, he led the American League with 231 hits, a club record that wasn't broken until 1986, when Don Mattingly had 238 hits. He was also an excellent bunter and outfielder. Earle Combs was elected to the Baseball Hall of Fame in 1970 by the Veterans Committee—the fourth-highest-ranked leadoff batter in the Hall of Fame.

Rickey Henderson played from 1979 to 2003—a very long 25-year career with nine different teams: 14 seasons (at four different times) with the Oakland Athletics, four seasons with the New York Yankees, three seasons with the San Diego Padres, and shorter stays with six other National and American League teams. Left fielder Henderson was a rarity in baseball—a right-handed batter who threw left-handed. He and Jimmy Ryan were the only two such players in this survey of 88 leadoff batters. Henderson led the American League in stolen bases 12 times, runs five times, walks four times, and hits and on-base percentage once each. Henderson is the all-time leader in stolen bases and runs and ranks second all-time in walks. He was selected to play in 10 All-Star Games and won one Gold Glove Award. Rickey Henderson was elected to the Baseball Hall of Fame in 2008, his first year of eligibility.

Dom DiMaggio played from 1940 to 1953: all 11 of his seasons were with the Boston Red Sox. He was the youngest of the three DiMaggio brothers, the other two being Vince and the legendary Joe. Dom was nicknamed the "Little Professor" because he was small (5 feet 9 inches and 168 pounds) and wore glasses. He was an excellent center fielder and a great leadoff batter. Dom led the American League in runs twice and triples and stolen bases once each. He was chosen to play in the All-Star Game in 7 of his 10 years as a regular player. DiMaggio's statistics would have been even better had he not lost three years of prime playing time (when he was 25, 26, and 27 years old) because of World War II. Dom DiMaggio has never been elected to the Baseball Hall of Fame, a prime example of players who would have been elected had not their careers been shortened because they served their country in a time of need.

Lu Blue played from 1921 to 1933: seven seasons with the Detroit Tigers, followed by three seasons with the St. Louis Browns, two seasons with the Chicago White Sox, and one last one-plate-appearance season with the Brooklyn Dodgers. He was a switch-hitting first baseman, one of only four leadoff-hitting first basemen in this 88-leadoff-batter survey. Blue did not have great numbers—he never led the American League in a major batting event—but he did play well throughout his career. For his 12 years as a regular player, he averaged 141 hits, 91 walks, and 96 runs. Blue had 150 or more hits, 90 or more walks, and scored 100 or more runs six times each. Remember, this was a time when teams played only 154 games each year. He was also a good-fielding first baseman. Lu Blue has never been elected to the Baseball Hall of Fame.

Stan Hack played from 1932 to 1947, spending all 16 of his seasons with the Chicago Cubs. He was the Cubs' third baseman and leadoff batter for most of those seasons. Not many third basemen have been leadoff batters—only 6 of the 88 leadoff batters in this study were third basemen. Hack was a very good leadoff batter: he led the National League in hits and stolen bases twice each and had a lifetime batting average of .301 and on-base percentage of .394. He had 75 or more walks in 9 of his 13 seasons as a regular player. In one seven-year stretch, Hack averaged 181 hits, including more than 190 hits three times. He was also one of the best-fielding third basemen of his time. Stan Hack has never been elected to the Baseball Hall of Fame.

George Burns played from 1911 to 1925: his first 11 seasons with the New York Giants, followed by three seasons with the Cincinnati Reds and one season with the Philadelphia Phillies. George Joseph Burns should not be confused with George Tioga Burns, who played in the American League from 1914 to 1929. George Joseph Burns was an excellent leadoff batter. He led the National League in walks and runs five times each, stolen bases twice, and on-base percentage once. He was also an excellent left fielder—Giant fans referred to the left-field stands as "Burnsville." George Burns has never been elected to the Baseball Hall of Fame.

Craig Biggio played from 1988 to 2007: all 20 of his seasons were with the Houston Astros. The first four seasons he was the catcher, the next 11 seasons he was the second baseman, the next two seasons he played in the outfield, and the last three seasons he was back at second base. Biggio was a slugging leadoff batter—he had a slugging average of more than .450 in nine of his seasons, and his 291 career home runs rank third all-time for leadoff batters. He led the league in doubles three times, runs twice, and stolen bases once. When Biggio batted, he wore an arm pad for protection because he crowded the plate and was susceptible to being hit with the ball. He led the league in hit by pitch five times, and his 285 career times hit by pitch ranks a close second all-time to the 287 for Hugh Jennings. Biggio was an excellent-fielding second baseman, frequently leading the National League in various defensive categories and earning him four Gold Glove Awards. He was selected to play in seven All-Star Games. Craig Biggio is not yet eligible for election to the Baseball Hall of Fame because he retired in 2007.

Juan Samuel played from 1983 to 1998, spending 16 seasons with seven different teams. He spent seven-plus seasons with the Philadelphia Phillies, two-plus seasons with the Los Angeles Dodgers, a year with the Cincinnati Reds, two seasons with the Detroit Tigers, and finally three seasons with the Toronto Blue Jays. At various times Samuel also played with the New York Mets and the Kansas City Athletics. He was a highly regarded second baseman blessed with both power and speed. Samuel became the first player ever to have 10 or more doubles, triples, home runs, and stolen bases in each of his first four years. He failed to realize his leadoff potential, however, because his high strikeout-to-walk ratio depreciated his speed. Samuel led the National League in triples twice but never led his league in any other major hitting event. Juan Samuel was selected to play in two All-Star Games but has never been elected to the Baseball Hall of Fame.

Pee Wee Reese played from 1940 to 1958: he spent his entire 16-year major league career with the Brooklyn / Los Angeles Dodgers. From 1943 to 1945 he served in the United States Navy. Reese was a right-handed-hitting shortstop. He was small (5 feet 9 inches and 176 pounds), as his nickname indicates, but he hit more for power than for average. From 1941 to 1956 he was in his prime—for 13 years he averaged 148 games, 152 hits, 94 runs, 85 walks, and 24 doubles. For the first 10 of those years, he was selected to play in the annual All-Star Game. Reese led the National League in runs, walks, and stolen bases once each. In 1984, the Veterans Committee elected Reese to the Baseball Hall of Fame.

The Number 2 Batter in the Order

The number 2 position in the batting order is often overlooked, but it is one of the most important of all. The second batter has fewer plate appearances with runners on base than the middle-of-the-order batter, but more plate appearances with runners on base than the leadoff batter. There are two important no-no's for second batters in the order: striking out is bad, but grounding into a double play is worse. Number 2 batters are a special category of player whose statistics should be compiled, analyzed, and judged apart from those who precede and succeed them in the batting order. It is difficult to identify number 2 batters in the baseball literature. The short list that has been compiled here is certainly not comprehensive or definitive, but it does constitute a start on a subject that needs to be addressed. Another difficulty in dealing with number 2 batters—and with leadoff batters as well—is that they often change positions in the batting order. They may bat second for most of their careers but bat leadoff or at some other position in the batting order at other times. The 2009 New York Yankees' switch of Derek Jeter from number 2 to leadoff and of Johnny Damon from leadoff to number 2 is a good example.

Incidentally, one of the leading number 2 batters of the Dead Ball Era, Germany Schaefer, was also one of the funniest players ever to play the game. Two of the funniest stories about Schaefer concern his hilarious reaction after delivering on his dramatic ninth-inning promise to hit a game-winning home run and his successful steal of first base. Yes, you read correctly. It was first base. These stories are re-counted by Lawrence S. Ritter in *The Glory of Their Times*, one of the greatest all-time books on baseball, which is highly recommended for your bookshelf.

The top 10 number 2 batters for whom references could be found in the baseball literature are identified in table 10.3. The PRG column is based on opportunity factors higher than used for leadoff batters and lower than used for middle-of-the-order batters. Interestingly, all 10 players were middle infielders: five were shortstops and five were second basemen. Shortstop Derek Jeter and second baseman Joe Morgan clearly dominate the list. Jeter is ranked number 1 in six and Morgan is ranked number 1 or number 2 in eight of the nine measures. Shortstops Alan Trammell and Johnny Pesky and second baseman Mark Loretta are the closest runners-up. Pesky ranks in the top three in seven measures, Trammell in the top four in eight measures, and Loretta in the top five in eight measures. From the modular perspective the rankings are quite cohesive. PRG and SLG agree on the exact ranking of the top 5 players, and OPS and LWTS agree on the exact ranking of 8 of the 10 players. Six of the measures rank the same five players in the top five, and the other three measures rank four of those players in the top five.

Table 10.4 summarizes the normalized PRG ratings for the top 10 number 2 batters. Normalization results in only minor changes in the ranking. Joe Morgan and Derek Jeter change positions, as do Johnny Pesky and Mark Loretta, whereas Alan Trammell retains his number 3 ranking. Germany Schaefer and Donnie Bush, both of whom played in the Dead Ball Era, gain three positions in the ranking. Mark Koenig, David Eckstein, and Jerry Remy, who played in live ball eras, lost two positions each in the rankings.

Before ending this section, we should say a few words about some of the leading number 2 batters. The current generation of baseball fans knows Joe Morgan as the knowledgeable television announcer for nationally televised baseball games. His reputation as a great player was established over a long 22-year career (1963–1984): 10 years with the Houston Astros, eight years with the Cincinnati Reds, and four years split with three other teams. Morgan was small (5 feet

TABLE 10.3 The Top 10 Number 2 Batters of All Time According to Various Measures

Player	PRG	Advanced Weighted Measures			Basic Weighted Measures			Unweighted Measures	
		LSLR	RC/27	LWTS	TA	OPS	SLG	OBP	AVG
Derek Jeter	1	1	2	1	2	1	1	3	1
Joe Morgan	2	2	1	2	1	2	2	2	8
Alan Trammell	3	4	4	4	4	4	3	6	4
Mark Loretta	4	8	5	5	5	5	4	4	3
Johnny Pesky	5	3	3	3	3	3	5	1	2
Mark Koenig	6	9	8	9	9	9	8	10	6
David Eckstein	7	5	7	6	7	6	6	7	5
Jerry Remy	8	7	10	10	10	8	7	8	7
Germany Schaefer	9	10	9	8	8	10	10	9	9
Donnie Bush	10	6	6	7	6	7	9	5	10

TABLE 10.4 Normalizing the PRG Ratings of the Top 10 Number 2 Batters

Before Normalization			Normalization		After Normalization		
Rank	Rating	Player	LG PRG	NORM[1]	Rank	Rating	Player
1	1.020	Derek Jeter	.985	1.036	1	1.144	Joe Morgan*
2	.991	Joe Morgan	.866	1.144	2	1.036	Derek Jeter
3	.945	Alan Trammell	.926	1.021	3	1.021	Alan Trammell
4	.910	Mark Loretta	.948	.960	4	.993	Johnny Pesky
5	.904	Johnny Pesky	.910	.993	5	.960	Mark Loretta
6	.831	Mark Koenig	.940	.884	6	.954	Germany Schaefer
7	.816	David Eckstein	.949	.860	7	.896	Donnie Bush
8	.765	Jerry Remy	.909	.842	8	.884	Mark Koenig
9	.745	Germany Schaefer	.781	.954	9	.860	David Eckstein
10	.744	Donnie Bush	.830	.896	10	.842	Jerry Remy

* Indicates that a player was elected to the Baseball Hall of Fame.

[1] Player's PRG rating divided by his league's PRG rating for the years in which he played.

7 inches and 160 pounds) but packed a punch at the plate. He led the National League in walks and on-base percentage four times each and in triples, slugging average, and runs once each. While with Cincinnati, he received two consecutive Most Valuable Player Awards and five consecutive Gold Glove Awards. Morgan was selected to play in 10 All-Star Games and is the only number 2 batter elected to the Baseball Hall of Fame. (See profile in chapter 6.)

Derek Jeter is the current shortstop for the New York Yankees. He has spent his entire 15-year career with the Yankees. Jeter has many of the assets desired for a number 2 hitter, including a high on-base percentage, the ability to steal bases, and the ability to avoid hitting into double plays. His one weakness, a relatively high strikeout rate (once every 5.91 at bats), is offset by his hitting credentials—a career batting average of .317 and slugging average of .459. He is also an excellent-fielding shortstop. In the 2009 season, Jeter was switched with Johnny Damon in the Yankees' batting order—Jeter from second to first and Damon from first to second. Jeter is already being extolled as the greatest shortstop in Yankee history and will probably be elected to the Baseball Hall of Fame after he retires.

Alan Trammell played from 1977 to 1996, spending all 20 seasons with the Detroit Tigers. As noted earlier in this chapter, Trammell and Lou Whitaker teamed together both in the field and at the plate. Second baseman Whitaker was the leadoff batter, and shortstop Trammell batted second. Trammell was an excellent fielder—he won four Gold Glove Awards—but he never led the American League in any major batting event. In the 15 years he was the regular shortstop, he averaged 143 hits, 25 doubles, and a batting average of .289. He batted over .300 seven times and had more than 30 doubles six times. Trammell grounded into few double plays, the primary no-no for number 2 batters. His best season was 1987, when he had 201 hits, including 34 doubles and 28 home runs, 105 runs batted in, a .343 batting average, an on-base percentage of .402, and a slugging average of .551. Trammell was selected to play in six All-Star Games but has never been elected to the Baseball Hall of Fame.

Johnny Pesky played from 1942 to 1954. He spent the first seven years of his career with the Boston Red Sox and divided the last three years between Boston, Detroit, and Washington. Pesky lost three years of prime playing time because of his military service in World War II. He led the American League in hits in 1942, when he returned from the war in 1946, and again in 1947, with over 200 hits each time. Pesky was also small (5 feet 9 inches and 168 pounds) but had a lifetime batting average of .307 and an on-base percentage of .394. He was fast, but not a base-stealing threat. Pesky had two other number 2 batter assets: low strikeout and grounding-into-double-plays percentages. He has served the Red Sox in many managerial roles after retiring as an active player and at 90 was still working in that capacity. Pesky is one of the most approachable and friendly human beings you will ever meet and one of the best representatives ever for the great game of baseball.

Mark Koenig played from 1925 to 1936. Koenig was the shortstop on the New York Yankees' Murderers' Row teams of the 1920s. He was one of the least murderous hitters on the team and one of only three everyday Yankee players not elected to the Hall of Fame. Koenig was, nevertheless, a good hitter—especially for a number 2 batting shortstop—and his poor fielding may have contributed more to his being passed over for the Hall of Fame. After leaving the Yankees, he had one good year. Koenig was instrumental in getting the Chicago Cubs to the World Series, but his old teammates, the Yankees, won the series in four straight games.

David Eckstein is still an active player. His statistics have declined in the last three years, but he is only 36 and could improve his career statistics enough to move ahead of Mark Koenig and Donnie Bush as the number 7 player.

* 11 *

The Table Clearers

The race for the distinction of being the greatest hitter of all time as measured by formal analysis is between Ruth and Williams, and it is so close that any revision of the method, however small, is likely to reverse the outcome.

The Bill James Historical Baseball Abstract, 1988, p. 394

....

The various measures of hitting performance rate and rank players in such a bewildering array of methods that their differences greatly exceed their similarities. We have found this to be true with regard to the hitters in each historical era and at each position played. We should not be surprised, therefore, if we find this to be also true with regard to all-time rankings, that is, for all hitters regardless of their historical era or position on the field. The top 25 middle-of-the-order hitters of all time are identified in table 11.1 according to the various measures we have been considering in the preceding chapters. It includes players from all eight historical eras with 4,000 or more plate appearances.

A total of 63 players share in the top 25 rankings in table 11.1 (including the footnotes). If you look at the top 25 players as a group, there is general agreement on 11 of them who belong in the top 25. All of the measures agree on five (Ruth, Gehrig, Williams, Pujols, and Hornsby), all except one measure agree on two (Foxx and Bonds), and all except two measures agree on five more (Greenberg, Brouthers, Helton, Musial, and Thomas). Twenty-four of the

TABLE 11.1 The Top 25 Middle-of-the-Order Batters of All Time According to Various Measures

Player	PRG	Advanced Weighted Measures			Basic Weighted Measures			Unweighted Measures	
		LSLR	RC/27	LWTS	TA	OPS	SLG	OBP	AVG
Babe Ruth	1	1	1	1	1	1	1	2	9
Lou Gehrig	2	2	5	3	4	3	3	5	15
Ted Williams	3	3	2	2	2	2	2	1	6
Hank Greenberg	4	7	15	9	11	7	7	—	—
Jimmie Foxx	5	11	11	7	6	6	5	11	—
Albert Pujols	6	4	17	6	8	4	4	13	21
Joe DiMaggio	7	9	24	19	—	13	10	—	—
Manny Ramirez	8	14	—	13	16	9	8	—	—
Barry Bonds	9	6	10	5	3	5	6	6	—
Dan Brouthers	10	10	4	10	17	—	—	16	8
Rogers Hornsby	11	20	16	12	9	8	11	9	2
Sam Thompson	12	15	23	—	—	—	—	—	—
Lance Berkman	13	16	—	22	19	16	23	—	—
Hack Wilson	14	—	—	—	—	—	25	—	—
Frank Thomas	15	17	—	18	18	15	24	21	—
Johnny Mize	16	—	—	—	—	20	17	—	—
Mark McGwire	17	—	20	24	15	11	9	—	—
Ed Delahanty	18	13	12	14	25	—	—	—	4
Al Simmons	19	—	—	—	—	—	—	—	20
Todd Helton	20	18	—	11	14	10	14	12	—
Alex Rodriguez	21	8	—	17	22	18	12	—	—
Albert Belle	22	—	—	—	—	—	16	—	—
Jim Thome	23	—	—	—	20	19	22	—	—
David Ortiz	24	—	—	—	—	—	—	—	—
Stan Musial	25	25	—	21	23	14	19	23	—
Larry Walker	(27)	—	19	—	21	17	15	—	—
Bill Joyce	(31)	12	7	16	10	—	—	7	—
Mel Ott	(32)	—	—	—	24	23	—	—	—
Jeff Bagwell	(33)	19	—	23	—	22	—	—	—
Vladimir Guerrero	(34)	22	—	—	—	21	13	—	—
Ty Cobb	(37)	23	—	15	13	—	—	8	1
Mickey Mantle	(38)	—	18	20	12	12	21	19	—

Continued on next page

Table 11.1 continued

Player	PRG	Advanced Weighted Measures			Basic Weighted Measures			Unweighted Measures	
		LSLR	RC/27	LWTS	TA	OPS	SLG	OBP	AVG
Chipper Jones	(41)	—	—	25	—	24	—	—	—
Joe Jackson	(46)	—	—	—	—	—	—	17	3
Tris Speaker	(70)	—	—	—	—	—	—	10	5
Pete Browning	(100+)	—	9	—	—	—	—	—	11
Billy Hamilton	(100+)	5	3	4	5	—	—	4	7
John McGraw	(100+)	21	6	8	7	—	—	3	22
Jesse Burkett	(100+)	25	—	—	—	—	—	25	16
Eddie Collins	(100+)	—	—	—	—	—	—	14	24
Other players[1]	—	—	—	—	—	—	—	—	—

[1] Twenty-three other players were ranked in the top 25 but in one measure only: Tip O'Neill #8, Roger Connor #13, Cap Anson #14, Harry Stovey #21, King Kelly #22, and Denny Lyons #25 in RC/27; Ralph Kiner #25 in OPS; Juan Gonzalez #18 and Willie Mays #20 in SLG; Ferris Fain #15, Max Bishop #18, Mickey Cochrane #20, Edgar Martinez #22, and Cupid Childs #24 in OBP; and Harry Heilmann #10, Willie Keeler #12, Bill Terry #13, George Sisler #14, Tony Gwynn #17, Nap Lajoie #18, Riggs Stephenson #19, Paul Waner #23, and Ichiro Suzuki #25 in AVG.

63 players share in the top 10 rankings, and 15 of them share in the top five rankings.

If you look at the specific rankings of the players within the top 25, the situation is the same as we have seen before: there is a lot of disagreement. Seven measures do agree that Ruth should be number 1, five measures agree that Williams should be number 2, and three measures agree that Gehrig should be number 3, Pujols should be number 4, Bonds should be number 6, Greenberg should be number 7, Brouthers should be number 10, and Foxx should be number 11. Nine players have two measures that agree on their specific ranking, and no measures agree on the specific ranking of 46 players.

This wide disarray presents the reader, who is looking for the names of the greatest hitters of all time, with a question that is very familiar by now: which measure should one choose? The answer, of course, was given in the Pregame Analysis, but this, in a sense, is the ultimate list of baseball hitters, so let's reiterate the framework for making the choice.

Your choice of a measure should be reasonable—one that is supported by objective evidence. It should not be an arbitrary, capricious, or nonsensical choice—one that is based, for example, on your favorite statistic or the statistic that ranks highest your favorite players or ranks highest those who played on your favorite team. Your choice should also be consistent—one that is the same for all times and for all positions played in the field.

Your choice should also consider the hierarchy of measurement groups in the table. Remember that this hierarchy was based on a combination of logical analysis and statistical testing. At the team level the PRG measure tested the very best of all, but it is not possible to subject the PRG player measure to the same test. The PRG player measure was, however, extrapolated from the PRG team measure, and the differences between the team and player measures are small.

Table 11.2 summarizes the normalized PRG ratings and rankings of the top 25 hitters of all time. It excludes players from the ECC and LBI for the reasons given in previous chapters. Those hitters who played most of their careers during the Dead Ball Era (Ty Cobb, Honus Wagner, Nap Lajoie, Sam Crawford, and Joe Jackson) and during the Dead Ball Interval (Willie Stargell, Dick Allen, Frank Robinson, and Hank Aaron) gained the most through normalization. Their hitting accomplishments were even greater than their raw statistics indicate because they were compiled in spite of the fact that pitchers had an advantage over batters when they played. Their raw statistics would have been much higher if they had played when hitters had the advantage over pitchers. Low tides lower all boats. Normalization adjusts for this inequity. Ten players who played most or all of their careers during the current Live Ball Enhanced Era (Manny Ramirez, Todd Helton, Frank Thomas, Mark McGwire, Alex Rodriguez, Albert Belle, Jim Thome, Carlos Delgado, David Ortiz, and Juan Gonzalez) lost the most through normalization. Their hitting was not as impressive as their raw numbers indicate, for they were compiled at a time when most hitter's statistics were higher because hitters had an advantage over pitchers. High tides raise all boats. Normalization adjusts for this advantage.

Babe Ruth appears to be the best hitter in the history of Major League Baseball—he ranks first in seven of the nine models before normalization and first in PRG after normalization. Lou Gehrig and Ted Williams are the closest runners-up. Before normalization, Williams ranks in the top four in all except one model and Gehrig in the top five in all except one model. Ruth, Gehrig, and Williams are the top three in five of the nine models, in the top four in six, and in the top five in eight models. After normalization, Williams ranks second and Gehrig third.

Many argue that Ted Williams would have had higher batting statistics than Babe Ruth if he hadn't been in the military service in World War II (three years) and the Korean conflict (nearly two years). He certainly had the potential for doing so—at the end of the 1942 season he had a career batting average of .356, including one season with an average of .406, one Most Valuable Player Award, and one Triple Crown to his credit. Williams would have been in his prime during World War II (the years he turned 25, 26, and 27) and would have had the potential to be one of the league's leading hitters during the Korean conflict (the years he turned 34 and 35). The year he turned 38 he led the league in OBP, the year he turned 39 he led the league in AVG, OBP, and SLG, and the year he turned 40 he led the league in AVG and OBP. There is, however, no way to know for sure what Williams's batting statistics would have been had he played Major League Baseball those five years.

The Elias Sports Bureau[1] once projected what Williams's statistics would have been had he played those five seasons. His total walks, runs, extra-base hits, and runs batted in would have been all-time major league records that would have remained to this day. His home runs would have been fewer than Ruth's, but the article speculated that Williams may have been tempted to play until he broke Ruth's record, much like Hank Aaron and Barry Bonds later did in order to set new all-time home run records.

Another way of looking at this issue is to calculate what Babe

TABLE 11.2 Normalizing the PRG Ratings of the Top 25 Middle-of-the-Order Batters of All Time

Before Normalization			Normalization		After Normalization		
Rank	Rating	Player	LG PRG	NORM[1]	Rank	Rating	Player
1	1.480	Babe Ruth	.933	1.586	1	1.586	Babe Ruth*
2	1.429	Lou Gehrig	.966	1.479	2	1.561	Ted Williams*
3	1.428	Ted Williams	.915	1.561	3	1.479	Lou Gehrig*
4	1.381	Hank Greenberg	.954	1.448	4	1.448	Hank Greenberg*
5	1.366	Jimmie Foxx	.959	1.424	5	1.442	Rogers Hornsby*
6	1.333	Albert Pujols	.931	1.427	6	1.428	Ty Cobb*[2]
7	1.326	Joe DiMaggio	.941	1.409	7	1.427	Albert Pujols
8	1.321	Manny Ramirez	.981	1.347	8	1.424	Jimmie Foxx*
9	1.291	Barry Bonds	.915	1.411	9	1.423	Joe Jackson[3]
10	1.283	Rogers Hornsby	.890	1.442	10	1.411	Barry Bonds
11	1.269	Lance Berkman	.940	1.350	11	1.409	Joe DiMaggio*
12	1.269	Hack Wilson	.955	1.329	12	1.407	Honus Wagner*[4]
13	1.268	Frank Thomas	.975	1.301	13	1.406	Johnny Mize*
14	1.265	Johnny Mize	.900	1.406	14	1.393	Nap Lajoie*[5]
15	1.265	Mark McGwire	.946	1.337	15	1.386	Mickey Mantle*[6]
16	1.253	Al Simmons	.963	1.301	16	1.380	Stan Musial*
17	1.251	Todd Helton	.950	1.317	17	1.364	Willie Stargell*[7]
18	1.248	Alex Rodriguez	.986	1.266	18	1.360	Dick Allen[8]
19	1.245	Albert Belle	.976	1.276	19	1.356	Hank Aaron*[9]
20	1.236	Jim Thome	.976	1.266	20	1.355	Sam Crawford*[10]
21	1.234	David Ortiz	.988	1.249	21	1.353	Frank Robinson*[11]
22	1.234	Stan Musial	.894	1.380	22	1.350	Lance Berkman
23	1.233	Harry Heilmann	.915	1.348	23	1.348	Harry Heilmann*
24	1.230	Larry Walker	.929	1.324	24	1.347	Mel Ott*[12]
25	1.230	Carlos Delgado	.974	1.263	25	1.347	Manny Ramirez

* Indicates that a player was elected to the Baseball Hall of Fame.

[1] Player's PRG rating divided by his league's PRG rating for the years in which he played.

[2] Before normalization Cobb was ranked 33rd with a PRG rating of 1.215 vs. a league average of .851.

[3] Before normalization Jackson was ranked 42nd with a PRG rating of 1.190 vs. a league average of .836.

[4] Before normalization Wagner was ranked 88th with a PRG rating of 1.140 vs. a league average of .810.

[5] Before normalization Lajoie was ranked 100+ with a PRG rating of 1.128 vs. a league average of .810.

[6] Before normalization Mantle was ranked 34th with a PRG rating of 1.214 vs. a league average of .876.

[7] Before normalization Stargell was ranked 59th with a PRG rating of 1.166 vs. a league average of .855.

[8] Before normalization Allen was ranked 66th with a PRG rating of 1.156 vs. a league average of .850.

[9] Before normalization Aaron was ranked 49th with a PRG rating of 1.178 vs. a league average of .869.

[10] Before normalization Crawford was ranked 100+ with a PRG rating of 1.085 vs. a league average of .803.

[11] Before normalization Robinson was ranked 55th with a PRG rating of 1.170 vs. a league average of .865.

[12] Before normalization Ott was ranked 29th with a PRG rating of 1.223 vs. a league average of .908.

Ruth's batting statistics would have been if he had been in the military service at the same ages that Williams was in the military service. Calculations indicate that he would have fallen short of Williams in most of the measures we have been talking about. Table 11.3 indicates that Ruth would have fallen short of Williams in PRG after normalization and in five of the other measures even before normalization. After normalization, he would have trailed Williams in two more measures as well (TA and OPS) because league batting statistics were higher during Ruth's years. Only in SLG would Ruth have exceeded Williams. Incredibly, Ruth would not have been a member of the 500-home-run club: his 480 home runs would have been 41 short of the 521 hit by Williams and far short of the number of home runs that would have been hit by Barry Bonds (614), Hank Aaron (569), and others under the same scenario. Thus, there is some evidence that Williams may actually have been a better hitter than Ruth. Williams never got "to walk down the street and have people say there goes the greatest hitter who ever lived." Surely, he must have hoped that serious students of the game might someday make the case for him. The continuing rivalry between their respective teams, the Yankees and the Red Sox, and their fans ensures that it will probably be a long time, if ever, before the final word is said on this subject.

Twenty of the 25 players in the after-normalization columns of table 11.2 were elected to the Baseball Hall of Fame. Since four of those players—Barry Bonds, Manny Ramirez, Albert Pujols, and Lance Berkman—are not eligible (Bonds because he stopped playing just three years ago and the others because they are still playing),

TABLE 11.3 The Babe versus The Kid: What Might Have Been (WMHB)

Player	PRG[1]	Advanced Weighted Measures			Basic Weighted Measures			Unweighted Measures	
		LSLR	RC/27	LWTS	TA	OPS	SLG	OBP	AVG
Ruth-actual	1.586	1.253	11.74	.608	1.460	1.164	.690	.474	.342
Williams-actual	1.561	1.202	11.50	.582	1.320	1.116	.634	.482	.344
Williams-actual	1.561	1.202	11.50	.582	1.320	1.116	.634	.482	.344
Ruth-WMHB	1.551	1.191	11.39	.558	1.346	1.132	.662	.470	.337

[1] PRG normalized.

only Dick Allen, among the top 25 hitters, has been passed over in elections to the Hall of Fame. Among the next 25 players, 12 were elected to the Hall of Fame, eight are active players or recently retired, and five have been passed over. Thus, of the top 50 hitters of all time, 38 were eligible for election and 32 (84%) were elected to the Hall of Fame.

Only 6 of those 38 eligible players have been passed over. Great hitting may not guarantee election to the Hall of Fame, but it almost does if you avoid certain things. Two of the six—Dick Allen and Albert Belle—were controversial players and may have been passed over for that reason. Mark McGwire was definitely passed over because of his involvement in the steroids controversy. Bob Johnson suffered because he always played on second-division teams. First basemen Dolph Camilli and Rudy York had to compete with four of the greatest first basemen ever—Lou Gehrig, Hank Greenberg, Jimmie Foxx, and Johnny Mize—and had to run against them in balloting for the Hall of Fame. Thus, if great hitters behave normally, avoid controversy, play on first-division teams, and don't have to compete with all-time greats at their position, they will probably be elected to the Hall of Fame.

PART III

Hot Stove League Favorites Revisited

* 12 *

Left on Base

It has always been my habit when I arrive at a restaurant kitchen to check
. . . to see if anything usable has been discarded and to inspect the walk-in
refrigerator to see if there are leftovers that can be recycled in one way
or another.

Jacques Pepin, *The Apprentice: My Life in the Kitchen,* p. 279

There are a number of baseball leftovers that need to be recycled or
disposed of before this work can be considered finished. They include the adjustment of baseball statistics to account for differences
in the configuration of ball parks (park effects), the performance of
players in critical situations (clutch hitting), the length of sustained
play (streakiness), and the way statistics can be calculated (a problem
with OPS).

Park Effects
Almost all major league teams, year in and year out, win more games
at home than on the road. In the 2009 season, for example, only 3
of the 30 major league teams failed to win more games at home than
on the road. In the 2007 and 2008 seasons, only 1 of the 30 major
league teams failed to win more games at home than on the road.
The home-field advantage in Major League Baseball is real, but it
is less than the home-field advantage in the NFL, NHL, and especially the NBA. Since the dimensions of football fields, hockey rinks,
and basketball courts are more or less uniform, the home-field ad-

vantage in those sports must be attributable to the familiarity of the surroundings and the vocal support of the home-team fans. Surely, the familiar surroundings and the vocal support of home-team fans must also be factors in baseball teams winning more games at home than on the road.

The home-field advantage in baseball is often attributed primarily to the crafting of baseball rosters to the peculiarities of ballparks. There are three primary differences between baseball stadiums that bear on home-field advantage: the surface (grass vs. artificial turf), the height and distance of fences from home plate, and the amount of foul territory. Big stadiums with artificial turf call for an emphasis on speed and defense. Long drives to the outfield are turned into fly outs, and line drives in the outfield gaps are turned into extra-base hits. Older stadiums with natural grass and shorter fences call for an emphasis on power. Teams with average-sized parks can safely tailor their rosters to their parks because most of their games on the road will be played in parks similar to their own. Batters can groove their swings to one kind of park and not have to change for a different kind of park very often. Batters on teams with unique parks, such as Fenway Park or Wrigley Field, have to change their approach in many, if not most, of the other parks they play in. For them, there may be a home-field advantage, but there is also a disadvantage on the road. The impact of the third factor in stadium differences—the amount of foul territory—is probably overstated. In a 1991 survey, the Elias Sports Bureau found that the amount of foul territory did have an effect on batting averages, but "the difference in foul outs between the most extreme stadiums is still only about one for both teams per game."[2]

This book does not factor park effects into the batting statistics for the three peculiarities in baseball stadiums because the home-field advantage of teams with unusually configured stadiums turns into a disadvantage when those teams play in stadiums on the road. A fourth difference in stadiums—the altitude of the stadiums themselves—is significant for only one team, the Colorado Rockies, who play their home games in the mile-high city of Denver, Colorado. The rarefied air of Denver definitely has a positive impact on hitting. This book, therefore, has factored park effects into the normalized batting statistics of Colorado Rockies players—notably Larry Walker and Todd Helton.

Clutch Hitting

Clutch hitting is dramatic, so dramatic in fact that a player's reputation for clutch hitting sometimes lingers longer than is justifiable. The concept of clutch hitting is plain and simple: getting a decisive hit at a critical time, which often determines the outcome of a game. Sharpening this concept into a concrete definition, however, is not so simple because it involves the consideration of four elements: the closeness of the score, the lateness of the game, the number of runners on base, and the number of outs. And each of these elements involves options: a game with a one-, two-, or three-run differential; from the seventh, eighth, or ninth inning; one, two, or three runners on base and which bases they are on; and one or two outs. Combinations of these elements and options are often called late-inning pressure situations.

The choice of elements and options to define clutch-hitting situations is completely subjective—there is no science involved. Whichever situation is chosen, however, the outcome is the same: successful clutch hitters drive in runs, especially with home runs. When hitters repeatedly do well in clutch situations, they become known as clutch hitters. The major leagues actually once used game-winning runs batted in as an official statistic but discontinued it after a brief trial.

Virtually all studies of clutch hitting are for limited periods of time—for a season, postseason, or consecutive seasons, but never for a player's entire career. All players with at least average-length careers go through a life cycle of performance—youth, maturity, and old age—whereby they learn, exploit, and lose their batting skills. Over the course of an entire career there are no clutch hitters—most players have their moments of successful clutch hitting, but they also have moments of unsuccessful clutch hitting.

A career clutch hitter, if he could be identified, would be a flop in the early innings of a game and/or in unimportant games. The team that scores the first run in a game generally wins the game, so it is perhaps more important to hit early in the game than it is to hit late in the game! The statistics of hitters who do both over the course of a career would be off the charts. Likewise, clutch hitters might hit well in important games but not in unimportant games; however, over the course of each season all games are equally important—a win is a win and a loss is a loss.

Clutch hitting, if it were to occur, would be reflected in higher-than-average runs batted in statistics (especially runs batted in from the bases). The statistic of runs batted in from the bases is a discreet element in the BPPA formula. Whatever clutch hitting a player has done over the course of his career is averaged with whatever non-clutch hitting he has done and is reflected in his BPPA and PRG statistics. Runs batted in is not a discreet element in any of the other measurements of hitting performance.

Streakiness

A subject closely related to clutch hitting is streakiness. Baseball players do not have uniform careers. They generally start at a modest level, thrive during their prime years, and slowly decline until they retire. All players, even the very best, have good seasons and bad seasons, and within each season players have streaks, periods of steady play, and slumps. Some players have more and/or longer streaks than slumps or vice versa. Players with long streaks in one season are likely to have shorter streaks in the next season. In the long run, there is no such thing as a streaky hitter or steady player—all players are a combination of both elements. They have recurring ups and downs, but with much less regularity than the movement of the tides and the rising and setting of the sun.

The concepts of streakiness and clutch hitting are intertwined, but streakiness and clutch-hitting situations are not necessarily coincidental. Sometimes they are and at other times they are not. When you're hot you're hot, and when you're not you're not. When you're streaking you're hitting in the clutch, and when you're slumping you're not hitting in the clutch. It's hard to imagine a streaking player who always fails in the clutch, or a slumping player who always hits in the clutch, or a steady hitter who either always hits in the clutch or never hits in the clutch. Thus, streaks, slumps, and clutch-hitting situations over time, and especially for an entire career, must be random happenings. They are intertwined like the warp and woof of a fabric, but the colors of the threads do not form a discernible pattern.

The Problems with On-Base Plus Slugging

The concept of the on-base plus slugging statistic reminds one of the old saying that two wrongs do not make a right. The weaknesses of each of the two halves of OPS may be countered by the strengths of the other half, but that doesn't mean that the two halves balance and completely offset each other. The weakness of OBP is that it does not weight batting events, and the weakness of SLG is that it does not include walks either as a time on base or as a batter opportunity. SLG does counter the OBP weakness by weighting hits, and OBP does counter the SLG weakness by including walks, but only in their own separate calculations, which are then simply added together. To truly integrate the calculation, walks and weighted hits should be added together and divided by hits, walks, and hit by pitch. If this were done, the OPS ratings and rankings of players would be different.

OPS has another problem: the unequal application of weights to players. Table 12.1 demonstrates how to calculate the value of each batting event in a player's career OPS rating—in this example the career of Ralph Kiner. The value of each batting event in the careers of five other great hitters was also calculated, and the results are summarized in table 12.2. The value of every single batting event is different for every player. Each player has a unique weight for each of his batting events. No two player batting events have the same weight. This discrepancy is masked by the fact that OPS is calculated in two separate sets.

The first OPS problem can be solved, as noted above, by integrating the OPS calculation with a common divisor. The second problem, however, is more difficult. It is obvious that the batting event weights should be the same for all players, but it is not so obvious which set of weights should be used. The arbitrary selection of a set

TABLE 12.1 Calculating the Value of Batting Events for Ralph Kiner

	BB/HP	1B	2B	3B	HR	Total
Number of events	1,035	827	216	39	369	2,486
Share of OBP	.166	.132	.035	.006	.059	.398[1]
Number of events weighted		827	432	117	1,476	2,852
Share of SLG		.159	.083	.022	.284	.548[2]
Share of OPS[3]	.166	.291	.118	.028	.343	.946[4]
Value of each event[5]	.000160	.000352	.000546	.000718	.000930	

[1] 2,486 divided by 6,240 (AB+BB/HP) = .398, and .166 + .132 + .035 + .006 + .059 = .398.

[2] 2,852 divided by 5,205 (AB) = .548, and .159 + .083 + .022 + .284 = .548.

[3] Share of OBP plus share of SLG.

[4] .398 + .548 = .946, and .166 + .291 + .118 + .028 + .343 = .946.

[5] Share of event divided by number of events (e.g., .166 ÷ 1,035 = .000160).

TABLE 12.2 The Value of OPS Batting Events in the Careers of Several All-Time Greats

Player	BB/HP	1B	2B	3B	HR	AB+B/HP
Ralph Kiner	.000160	.000352	.000546	.000718	.000930	6,240
Rogers Hornsby	.000108	.000231	.000351	.000473	.000598	9,259
Mickey Mantle	.000101	.000226	.000349	.000472	.000595	9,848
Al Kaline	.000087	.000187	.000283	.000386	.000484	11,449
Tris Speaker	.000086	.000184	.000282	.000378	.000479	11,679
Carl Yastrzemski	.000072	.000155	.000240	.000322	.000407	13,873

of weights could help some players and hurt others. The determination of which set of weights to use should be based on a comprehensive study of perhaps hundreds of players, which is beyond the scope of this book. In the meantime, we will have to content ourselves with the admonition—reader beware—handle OPS with care.

The use of OPS to rate players for a single season is also suspect. Table 12.3 uses the hypothetical example of two players with virtually the same OBP, SLG, and OPS averages, but with widely varying numbers of plate appearances. This is not an unrealistic example because with 3.1 plate appearances for every scheduled game, the player with the fewest plate appearances would still qualify for the batting average and slugging average titles (3.1 × 162 = 502).

Player A, the one with 650 plate appearances, in effect, has lower weights applied to his batting events than player B, the one with

TABLE 12.3 Comparing the Value of OPS Batting Events for Two Hypothetical Players in One Season

	BB/HBP	1B	2B	3B	HR	AB	PA	OBP	SLG	OPS
PLAYER A	100	75	40	5	40	550	650	.400	.600	1.000
Share of OBP	.154	.114	.062	.008	.062			.400		
Share of SLG		.137	.145	.027	.291				.600	
Share of OPS	.154	.251	.207	.035	.353					1.000
Value per event	.001540	.003347	.005175	.007000	.008825					
PLAYER B	75	61	30	5	30	427	502	.401	.600	1.001
Share of OBP	.149	.122	.060	.010	.060			.401		
Share of SLG		.143	.141	.035	.281				.600	
Share of OPS	.149	.265	.201	.045	.341					1.001
Value per event	.001987	.004344	.006700	.009000	.011367					
Difference in VPE[1]	.000447	.000997	.001525	.002000	.002542					
% of A to B[2]	77.5	77.0	77.2	77.8	77.6		77.2			

[1] Difference in value per event.

[2] For PA column % of B to A.

only 502 plate appearances. The weights applied to player A's batting events are about 75% of the weights applied to player B's batting events. The weights applied to the career player with the most plate appearances in table 12.2 are only 45% of those applied to the player with the least plate appearances. Thus, the differences in weights applied to players are higher with regard to career statistics than with regard to seasonal statistics. Seasonal differences exist, however, and constitute a potential problem, the extent of which depends on player differences in the number of plate appearances.

* 13 *

Whatever Happened to the Triple Crown?

Hitting a baseball is the single most difficult thing to do in sport. . . . A hitter . . . is expected to hit a round ball with a round bat and adjust his swing in a split second to 100-mile-per-hour fastballs, backbreaking curveballs, and, occasionally, knuckleballs that mimic the flight patterns of nearsighted moths. . . . Even the vaunted major leaguer who hits at the magic .300 level . . . fails seven times every ten at bats.

Ted Williams, *Ted Williams' Hit List,* pp. 19 and 21

It has been 43 years since any baseball player has won the Triple Crown of baseball (leading the league in batting average, home runs, and runs batted in). Carl Yastrzemski, of the 1967 "impossible dream" Boston Red Sox, was the last player to win the Triple Crown. It was the ninth time it had been done in the previous 41 years, for an average of about once every four and a half years. The times have changed, and the Triple Crown now seems like another impossible dream.

Table 13.1 lists the players who have won the Triple Crown of baseball, and table 13.2 expands this list into a Triple Crown hierarchy. It has always been difficult to win the Triple Crown. In the entire 134-year history of the National League, it has been done only five times—for an average of once every 26.8 years. It hasn't been quite as difficult in the American League, as it has been accomplished nine times in 109 years—for an average of once every 12.1 years. Some of the greatest players of all time never won a Triple Crown—Babe

Ruth, Hank Aaron, Willie Mays, Joe DiMaggio, and Barry Bonds, to name just a few. Rogers Hornsby and Ted Williams were the only two players to win two Triple Crowns. Only once, in 1933, was the Triple Crown won in both leagues in the same year—Jimmie Foxx in the American League and Chuck Klein in the National League. It was a truly unique happening, as they both played for teams from the same city: Foxx played for the Philadelphia Athletics, and Klein played for the Philadelphia Phillies. Ty Cobb was the youngest player to win the Triple Crown (a few months shy of his 23rd birthday), and Lou Gehrig was the oldest (three months after his 31st birthday).

Why is it so difficult to win the Triple Crown? And why has it seemed almost impossible in recent years? The answer to the first question is no great secret. Almost anyone who has played baseball knows the answer. Just hitting a baseball is difficult enough, as Ted Williams so articulately explained above. To win the Triple Crown of baseball, you have to combine hitting for strength (home runs) with hitting with skill (batting average) and do it in a timely fashion,

TABLE 13.1 The Triple Crown of Baseball

Year	Winner	Team	League
1878	Paul Hines	Providence Grays	National
1887	Tip O'Neill	St. Louis Cardinals	American Assoc.
1901	Nap Lajoie	Philadelphia Athletics	American
1909	Ty Cobb[1]	Detroit Tigers	American
1922	Rogers Hornsby	St. Louis Cardinals	National
1925	Rogers Hornsby[1]	St. Louis Cardinals	National
1933	Jimmie Foxx	Philadelphia Athletics	American
1933	Chuck Klein	Philadelphia Phillies	National
1934	Lou Gehrig[1]	New York Yankees	American
1937	Joe Medwick	St. Louis Cardinals	National
1942	Ted Williams[1]	Boston Red Sox	American
1947	Ted Williams	Boston Red Sox	American
1956	Mickey Mantle[1]	New York Yankees	American
1966	Frank Robinson	Baltimore Orioles	American
1967	Carl Yastrzemski	Boston Red Sox	American

[1] Ty Cobb, Rogers Hornsby, Lou Gehrig, Ted Williams, and Mickey Mantle constitute an elite of the elites: in these years they led both leagues in batting average, home runs, and runs batted in.

TABLE 13.2 The Triple Crown Hierarchy

A. Players who led both leagues in BA, HR, & RBI	
1. Ty Cobb (1909)	4. Ted Williams (1942)
2. Rogers Hornsby (1925)	5. Mickey Mantle (1956)
3. Lou Gehrig (1934)	

B. Players who won the Triple Crown in their league only	
1. Paul Hines (1878)	6. Chuck Klein (1933)
2. Tip O'Neill (1887)	7. Joe Medwick (1937)
3. Nap Lajoie (1901)	8. Ted Williams (1947)
4. Rogers Hornsby (1922)	9. Frank Robinson (1966)
5. Jimmie Foxx (1933)	10. Carl Yastrzemski (1967)

C. Players who narrowly missed winning the Triple Crown[1]	
1. John Reilly (1884)	7. Rogers Hornsby (1921 and 1924)
2. Hugh Duffy (1894)	8. Babe Ruth (1923, 1924, and 1926)
3. Cy Seymour (1905)	9. Jimmie Foxx (1932 and 1938)
4. Ty Cobb (1907 and 1911)	10. Ted Williams (1949)
5. Honus Wagner (1908)	11. Al Rosen (1953)
6. Gavvy Cravath (1913)	

[1] First in two events and second in one event.

that is, with runners on base (runs batted in). Triple Crown hitting is three times as difficult as ordinary hitting. It takes three very special talents to win the Triple Crown. You also have to have a little luck because runs batted in are from the bases and you therefore have to have teammates on base when you come to bat.

The answer to the second question is found partly in the nature of our times. This is an age of specialization, and baseball is no different from other activities. Baseball players emphasize either the home run or the batting average, but not both. Power hitting has become the name of the game. Twenty-six of the 50 all-time leaders in career home run percentage (with 1,000 or more games) are playing or played during the current Live Ball Enhanced Era, and eight of them ranked in the top 13 (Barry Bonds, Jim Thome, Adam Dunn, Alex Rodriguez, Albert Pujols, Sammy Sosa, Juan Gonzalez, and Manny Ramirez). Only 4 of the 50 all-time leaders in batting average (with 1,000 or more games) are playing in the current Live Ball Enhanced era (Albert Pujols, Ichiro Suzuki, Todd Helton, and

Vladimir Guerrero), and none of them rank in the top 10. Albert Pujols and Vladimir Guerrero are the only players active in the current Live Ball Enhanced Era who rank in the top 50 in both home run percentage and batting average.

Perhaps the biggest factor working against winning the Triple Crown is the expansion over a number of seasons from 8 to 14 teams in the American League and from 8 to 16 teams in the National League. The more competition there is, the more difficult it is for one player to lead their league in all three events, but it does not make it impossible. Frank Robinson and Carl Yastrzemski won the Triple Crown in a 10-team American League. Larry Walker (1997) and Barry Bonds (2002) narrowly missed leading the National League (with 14 teams in 1997 and 16 teams in 2002) in batting average and home runs, the most difficult of the Triple Crown events to win simultaneously, and almost as difficult as winning the Triple Crown itself. Possibly the most hopeful observation of all is that five Triple Crown winners led both leagues (16 teams at that time) in all three events.

Table 13.3 lists the players who have won the so-called "Double Crown" of baseball—that is, players who have won two legs of the Triple Crown but not the third leg. The Double Crown has been won 123 times—92 in home runs and runs batted in, 25 in runs batted in and batting average, but only 6 times in batting average and home runs. Thus, the most difficult Double Crown combination is batting average and home runs. In the 134 National League and 109 American League seasons, a player has led his league in batting average and home runs only 19 times (14 Triple Crowns and 5 Double Crowns), for an average of once every 16.8 years in the National League and once every 9.9 years in the American League. One Triple Crown was won by an American Association player (Tip O'Neill), and one Double Crown was won by a Union Association player (Fred Dunlap). Both of these leagues were regarded as major leagues at that time.

The players who have won the Triple Crown and those who have won or have come close to winning the Double Crown in batting average and home runs are listed in table 13.4. Ted Williams was the

TABLE 13.3 The "Double Crown" of Baseball

Year	Player	Team	League
A. Batting average and home runs			
1884	Fred Dunlap	St. Louis Maroons	Union Assoc.
1894	Hugh Duffy	Boston Braves	National
1912	Heinie Zimmerman	Chicago Cubs	National
1924	Babe Ruth	New York Yankees	American
1939	Johnny Mize	St. Louis Cardinals	National
1941	Ted Williams	Boston Red Sox	American
B. Batting average and runs batted in[1]			
1881 & 1888	Cap Anson	Chicago White Stockings	National
1883 & 1892	Dan Brouthers	Buffalo Bisons & Brooklyn Dodgers	National
1907, 1908 & 1911	Ty Cobb	Detroit Tigers	American
1908 & 1909	Honus Wagner	Pittsburgh Pirates	National
1920 & 1921	Rogers Hornsby	St. Louis Cardinals	National
C. Home runs and runs batted in[2]			
6 *times*	Babe Ruth	New York Yankees	American
4 *times*	Mike Schmidt	Philadelphia Phillies	National
3 *times*	Hank Greenberg	Detroit Tigers	American
3 *times*	Hank Aaron	Milwaukee Brewers & Atlanta Braves	National

[1] Fourteen players led their league in batting average and runs batted in once each: Deacon White, Sam Thompson, Ed Delahanty, Nap Lajoie, Cy Seymour, Sherry Magee, Paul Waner, Jimmie Foxx, Stan Musial, Tommy Davis, Joe Torre, Al Oliver, Todd Helton, and Matt Holiday.
[2] Fourteen other players led their league in home runs and runs batted in twice each: Ed Delahanty, Harry Davis, Home Run Baker, Gavvy Cravath, Bill Nicholson, Johnny Mize, Willie McCovey, Harmon Killebrew, Johnny Bench, George Foster, Jim Rice, Cecil Fielder, Alex Rodriguez, and Ryan Howard; 49 players did it once each.

only player in the entire history of Major League Baseball to lead his league in batting average and home runs 3 times (two Triple Crowns and one Double Crown), and he narrowly missed two other times—in 1949 (when he missed a third Triple Crown because he lost the batting title to George Kell .3428 to .3429) and in 1957. Rogers Hornsby led his league in batting average and home runs twice (by virtue of his two Triple Crowns), and 15 other players did it once. It hasn't been done in the National League since 1939 (Johnny Mize) and in the American League since 1967 (Carl Yastrzemski).

TABLE 13.4 Batting Average and Home Runs—A Jewel in Itself

Player	Triple Crown	Double Crown (HR & AVG)	Almost a Double Crown (HR & AVG)[1]	Total
Ted Williams	2	1	2	5
Rogers Hornsby	2	—	2	4
Jimmie Foxx	1	—	3	4
Babe Ruth	—	1	3	4
Ty Cobb	1	—	2	3
John Reilly	—	—	2	2
Lou Gehrig	1	—	—	1
Paul Hines	1	—	—	1
Chuck Klein	1	—	—	1
Nap Lajoie	1	—	—	1
Mickey Mantle	1	—	—	1
Joe Medwick	1	—	—	1
Tip O'Neill	1	—	—	1
Frank Robinson	1	—	—	1
Carl Yastrzemski	1	—	—	1
Hugh Duffy	—	1	—	1
Fred Dunlap	—	1	—	1
Johnny Mize	—	1	—	1
Heinie Zimmerman	—	1	—	1
Other players	—	—	17[2]	—

[1] First in one and second in the other event.

[2] Barry Bonds, Dan Brouthers, Pete Browning, Gavvy Cravath, Sam Crawford, George Hall, Tommy Holmes, Derek Lee, Fred Lynn, Willie Mays, Alex Rodriguez, Al Rosen, Jimmy Ryan, Cy Seymour, George Sisler, Honus Wagner, and Larry Walker.

It is interesting to look at the winners of the Triple Crown in relation to the positions they played. The position that has produced the most Triple Crowns is left field (five): Ted Williams (two), Joe Medwick, Tip O'Neill, and Carl Yastrzemski. Three center fielders (Ty Cobb, Paul Hines, and Mickey Mantle) and two right fielders (Chuck Klein and Frank Robinson) also won the Triple Crown. The infield has produced five Triple Crowns: Rogers Hornsby (two) and Nap Lajoie at second base, and Jimmie Foxx and Lou Gehrig at first base. No shortstop, third baseman, or catcher has ever won a Triple Crown.

It is also interesting to look at the teams that the winners of the Triple Crown played for. The St. Louis Cardinals had the most Triple Crown winners (Rogers Hornsby twice, Tip O'Neill, and Joe Medwick). The Boston Red Sox were second (Ted Williams twice and Carl Yastrzemski). The Philadelphia Athletics had two (Nap Lajoie and Jimmie Foxx), as did the New York Yankees (Lou Gehrig and Mickey Mantle). Four other teams had one each—the Philadelphia Phillies, the Detroit Tigers, the Baltimore Orioles, and the Providence Grays.

All except three Triple Crown winners were elected to the Baseball Hall of Fame. Fred Dunlap, Paul Hines, and Tip O'Neill played when the home run was rare and therefore not recognized as a major statistic. Paul Hines and Ty Cobb won the Triple Crown with only four and nine home runs, respectively. Only later when the home run became more common and therefore extolled was the Triple Crown recognized as a prize, whereupon Dunlap, Hines, and O'Neill became retroactive winners of the Triple Crown. The fact that John Reilly (a near Triple Crown winner) and O'Neill played most of their careers in the American Association may have been another factor in their being passed over for the Hall of Fame.

The probability of someone winning the Triple Crown in the near future is not great. Among active players, only six have even won a Double Crown—Todd Helton (2000), Alex Rodriguez (2002 and 2007), Andruw Jones (2005), Ryan Howard (2006 and 2008), David Ortiz (2006), and Matt Holliday (2007)—and none of them combined batting average and home runs or led the league in home runs and ranked as high as third in batting average. Someone will win the Triple Crown again sometime. An exceptional player will have an exceptional season, or less emphasis may come to be placed on strength and more on a balance of strength and skill. The talented player who combines strength, skill, and timeliness stands a much better chance for genuine baseball immortality than the one-dimensional power hitter who concentrates solely on hitting home runs.

The Triple Crown performances of the past were truly remarkable. Let's hope that we will witness such performances in the future. Be on the lookout for the player who has the potential for leading his league in both batting average and home runs. A player with the ability to combine hitting for accuracy with hitting for power will inevitably have a lot of runs batted in. Maybe the dream of a Triple Crown winner is not an impossible one after all.

In the meantime, each major league should give an annual Strength, Skill, and Timeliness (SST) Award. Only players ranking in the top five in batting average, home runs, and runs batted in or players ranking first in two of those events would qualify. For each of the three events, the league leader would get five points, the runners-up would get four points, and so on through the top five. The player with the most total points would win the SST Award. If this award had been in effect the past five years, Albert Pujols would have won the award three times and Matt Holliday and Ryan Howard once each (2007 and 2008, respectively) in the National League. Alex Rodriguez would have won the award twice and David Ortiz and Mark Teixeira once each (2006 and 2009, respectively) in the American League. No American League player would have qualified in 2008.

We live in an age of specialization. In the baseball world, batters try to concentrate on getting hits or hitting home runs. Players who do both are rare, and those who do both with runners on base are even rarer. Being able to do more than one of these things is a great talent. Any team would like to have such a player in its batting order. Giving an annual SST Award would be a great way to recognize players who come closest to the Triple Crown ideal.

Postgame Report

It's déjà vu all over again.

Yogi Berra with Dave Kaplan, *What Time Is It? You Mean Now?* p. 137

The term *new* in the subtitle of this book—A New Way to Rate Baseball Players—refers to more than just the new BPPA/PRG formula. It also refers to the fact that players are ranked differently depending on whether they batted leadoff, second, or in the middle of the batting order. A third new idea is the systematic comparison of the ways different measures rank players. A fourth new idea is that of a "Double Crown," which is batting average and home runs, and how rare it is for a player to win—almost as rare as winning the Triple Crown itself. A fifth new idea is that of a Strength, Skill, and Timeliness Award as a way to recognize players who come closest to the Triple Crown ideal.

The preface of this book ended with a question—can baseball fans make meaningful comparisons between players and interpret what takes place on the field without statistics?—and the answer to that question was a resounding no! This book is based on the underlying assumption that baseball cannot be understood without an analysis of the statistics with which it is inextricably intertwined. Trying to understand baseball without statistics is like trying to understand a 3-D movie without 3-D glasses.

Let's revisit the major findings of this book.

1. The history of Major League Baseball can be divided into eight distinct eras based on differences in the number of earned runs scored per game. The eras alternate back and forth—like the ocean tides and the rising and the setting of the sun—between those in which pitchers tend to dominate and those in which hitters tend to dominate.

2. There were so many changes during the first historical era (1876–1892) that it can appropriately be called the Era of Constant Change.

3. The performance of hitters is measured in many different ways. Baseball logic, supported by statistical testing, indicates that a new measure—potential runs per game (PRG)—is a reliable way to assess a hitter's performance.

4. Each of the eight historical eras of baseball had a clearly dominant overall hitter: Dan Brouthers in the Era of Constant Change, Ed Delahanty in the Live Ball Interval, Ty Cobb in the Dead Ball Era, Babe Ruth in the Live Ball Era, Ted Williams in the Live Ball Continued era, Hank Aaron in the Dead Ball Interval, and Mike Schmidt in the Live Ball Revived era. The identity of the dominant hitter in the current Live Ball Enhanced Era is uncertain, however, because of the steroids controversy.

5. The official rules of the game for the first two eras of baseball (1876–1900) were very different from what they were thereafter. There was also a very high rate of errors, and the information on some batting events is incomplete. Thus, it is impossible to make a fair comparison of hitters during those eras with hitters who played later.

6. The various defensive positions place different physical demands on players that influence their potential as hitters. The respective positions of players in the batting order also influence their potential as hitters.

7. Different performance standards should be applied to players depending on what position they play in the field and where they hit in the batting order.

8. The greatest middle-of-the-order hitters at each position in the field were as follows: Babe Ruth in right field, Ty Cobb in center field, Ted Williams in left field, Lou Gehrig at first base, Rogers Hornsby at second base, Honus Wagner at shortstop, Mike Schmidt at third base, and Roy Campanella as catcher.

9. Four players were ranked number 1 for the historical era in which they played and for the position that they played in the field. Babe Ruth was ranked number 1 for the Live Ball Era and for right field. Ted Williams was ranked number 1 for the Live Ball Continued era and for left field. Ty Cobb was ranked number 1 for the Dead Ball Era and for center field. And Mike Schmidt was ranked number 1 for the Live Ball Revived era and for third base.

10. The 10 greatest leadoff batters of all time are Bobby Bonds, Augie Galan, Rod Carew, Elbie Fletcher, Wade Boggs, Paul Molitor, Tim Raines, Topsy Hartsel, Pete Rose, and Lou Whitaker, in that order. Billy Hamilton, Jimmy Ryan, and Jesse Burkett may deserve to be in the top 10 instead of some of the players listed, but it is impossible to say for sure because the baseball rules were so different when they played compared with subsequent eras.

11. The 10 greatest number 2 batters of all time are Joe Morgan, Derek Jeter, Alan Trammell, Johnny Pesky, Mark Loretta, Germany Schaefer, Donnie Bush, Mark Koenig, David Eckstein, and Jerry Remy, in that order. Some other player or players may deserve to be in the top 10 instead of some of these players, but the baseball literature on number 2 players is so sparse that, at this time, it is impossible to say for sure. More work needs to be done on this subject.

12. The 10 greatest middle-of-the-order hitters (third, fourth, and fifth in the order) of all time are Babe Ruth, Ted Williams, Lou Gehrig, Hank Greenberg, Rogers Hornsby, Ty Cobb, Albert Pujols, Jimmie Foxx, Joe Jackson, and Barry Bonds, in that order. A case can be made that Ted Williams was actually a better hitter than Babe Ruth, but general agreement on this subject is difficult because of the intense rivalry between their respective teams, the Boston Red Sox and the New York Yankees. If we had more information on the

early years of Major League Baseball and Negro League Baseball, this lineup might have included players like Dan Brouthers and Ed Delahanty from the former and Josh Gibson and Buck Leonard from the latter.

13. In the entire history of Major League Baseball, the Triple Crown has been won only 15 times, and only two players—Rogers Hornsby and Ted Williams—have won two Triple Crowns. In the last 43 years, no player has won the Triple Crown because power hitting has become the name of the game.

14. The Triple Crown of baseball is based on the difficult combination of strength, skill, and timeliness. The pursuit of these three worthy objectives should be encouraged by granting an annual SST Award to the player with the best combined rating for batting average, home runs, and runs batted in.

15. The home-field advantage of extreme ballparks is more than offset by an away-from-home-field disadvantage for hitters on teams based in extreme ballparks.

16. There is no such thing as career clutch hitting. The normal pattern of hitting is an endless ebb and flow of streaks and slumps of varying lengths throughout a player's career.

The all-time PRG lists of players by position in the field and by position in the batting order are noted above. Combining these lists into one all-time "dream team" would be difficult because they are mutually exclusive. In the world of dreams, however, we have the luxury of picking any player in any position that occurs to our subconscious. My dream team would consist of leadoff batter center fielder Ty Cobb, second batter second baseman Rogers Hornsby, third batter right fielder Babe Ruth, fourth batter left fielder Ted Williams, fifth batter first baseman Lou Gehrig, sixth batter shortstop Honus Wagner, seventh batter third baseman Mike Schmidt, and eighth batter catcher Roy Campanella. If this dream team had to play in the American League after 1972, my designated hitter would be Hank Aaron.

And waiting in the dugout would be another "dream team" of backup players eager to enter the game at any moment. It would include left fielder Joe Jackson, center fielder Joe DiMaggio, right fielder Frank Robinson, third baseman Chipper Jones, second baseman Jackie Robinson, shortstop Ernie Banks, first baseman Hank Greenberg, catcher Mike Piazza, and designated hitter Albert Pujols. It is hoped that *The Runmakers* will encourage you to choose your own "dream team."

Appendix: Using the BPPA Formula in Fantasy Baseball Leagues

In most fantasy baseball leagues, the scoring of runs is based on a somewhat arbitrary combination of batting, pitching, and fielding statistics—AVG, HR, R, RBI, ERA, and FA, for example. This approach does not recognize the process by which runs are actually scored. In the real baseball world, the scoring of most runs requires teamwork—a batter gets on base, another batter advances him along, and a third batter drives him home. Only in the case of a solo home run is just one batter involved in scoring a run.

A more realistic fantasy league approach would be to use a kind of double-entry bookkeeping ledger for the scoring of runs—one for the scoring of a team's runs and another for the scoring of runs by a team's opponents. The BPPA formula would constitute a very realistic way of counting for the runs a team scores, and a combination of ERA and FA could be used to account for the runs scored by a team's opponents. The team with the biggest positive difference between the runs it scores and the runs scored by its opponents would be declared the winning team.

The BPPA formula could also be used by fantasy league managers to help them draft the best available players for their team rosters. This would not guarantee them the best available choices because a player's performance in prior years is not necessarily a good predictor of how he will perform in the coming year. Players have ups and downs with both good years and not so good years. The BPPA formula is based on player careers, which even out the good and not so good years. There are also ups and downs in other player statistics—AVG, OBP, SLG, etc. Predicting a batter's numbers in any statistic is not a cut-and-dried process. That's what makes baseball a game of many surprises. However, a reliable measure for explaining a prior year's performance, that is, BPPA, should also be a reliable measure for guessing a player's performance in the coming year.

It would not be necessary to wait until the end of the baseball season to total up the numbers and decide on the winner. The statistics could be totaled at regular intervals—let's say for April–May, June–July, and August–September, and, finally, for the entire season. This would help the participants maintain an active interest in the league throughout the year. The league could decide the amount to be awarded to the winners of the various periods. The biggest prize, presumably, would go to the winner for the entire year.

Notes

Pregame Analysis

1. Baseball Info Solutions and Bill James, *The Bill James Handbook 2008* (ACTA Sports, 2007), p. 477.

2. Jim Albert and Jay Bennett, *Curve Ball: Baseball, Statistics, and the Role of Chance in the Game* (New York: Springer, 2001).

3. Consider the following example:

	BPPA		PA per G		BPG		BPR		PRG
Player A	1.00	×	5	=	5.00	÷	4	=	1.250
Player B	.90	×	5	=	4.50	÷	4	=	1.125
Player C	.80	×	5	=	4.00	÷	4	=	1.000

Player B's BPPA is 90% of player A's, and player B's PRG is 90% of player A's (1.125 ÷ 1.250 = 90%). Similarly, player C's BPPA is 80% of player A's, and player C's PRG is 80% of player A's (1.000 ÷ 1.250 = 80%). Thus, the conversion process does not affect the relative rating and ranking of the players.

4. The number of earned runs per game was used—instead of total runs per game—because of the high rate of errors and, consequently, the high number of unearned runs in the early days of baseball. The error rate has declined era by era to the present time. In the first era of Major League Baseball, the Era of Constant Change, 43% of the total runs were unearned vs. only 8% of the total runs in the current era.

Chapter 1. The Era of Constant Change

1. Stefan Szymanski, "Baseball Economics," in *Handbook on the Economics of Sport*, ed. Wladimir Andreff and Stefan Szymanski (Northampton, Mass.: Edward Elgar Publishing, 2006); Harold Seymour, *Baseball: the Early Years* (Oxford: Oxford University Press, 1960); and Zev Chafets, *Cooperstown Confidential* (New York: Bloomsbury, 2009).

2. The American Association was formed in 1882 and succeeded in raiding the ros-

ters of the National League teams. In 1883 both leagues signed the National Agreement, which recognized both leagues as major leagues and included a reserve clause that bound players to their teams. The American Association and the National League competed with each other for 10 seasons, from 1882 to 1891. In 1882 the champions of each league split two exhibition contests. The following year the American Association team refused to play the National League team. From 1884 to 1890 the champions of each league played each other for what was billed as the World Championship. After the 1891 season no postseason games were held because the leagues quarreled and the American Association folded. The Union Association had been formed in 1884. It sought to attract players from the National League and American Association by denouncing the reserve clause, but it folded after the season. Another effort to defeat the reserve clause was attempted in 1890 when disgruntled players organized the Players League. It also folded after only one season.

Chapter 2. The Live Ball Interval

1. The same 12 cities had teams in the National League in every season from 1892 to 1899. In 1900 the League contracted to eight teams, and the following year three of the teams that left the National League joined the newly formed American League.

2. The pitcher's rubber was enlarged in 1895, but it wasn't until 1900 that home plate was enlarged from a 12-inch square to a five-sided figure 17 inches wide. This proved to be so favorable to pitchers over batters that it marks the end of one historical period and the beginning of another.

Chapter 3. The Dead Ball Era

Epigraph. As quoted on p. 60 of Lawrence S. Ritter's *The Glory of Their Times* (New York: HarperCollins, 1992).

1. It is ironic that the United States, the leading free-market country in the world, has continually restricted its market in order to perpetuate its national pastime. The reserve clause, cartel-like negotiations with the media, the luxury tax, and the rookie draft are among the anti-free-market restrictions employed by the major leagues and tolerated by the courts and the political branches of government. These market restrictions are used to maintain a nominally free market structure, but baseball fans and the general public seem completely unaware of or concerned about this irony.

Chapter 4. The Live Ball Era

1. Prior to the next season (1921), 17 pitchers (8 in the National League and 9 in the American League) were named spitball pitchers for the rest of their major league careers.

2. In the Live Ball Era, players hit more than 40 home runs 30 times and in none of these cases was the player's batting average below .300. Since then, the number of times a player hit more than 40 home runs and had a batting average below .300 was 14 in the Live Ball Continued era, 15 in the Dead Ball Interval, and 15 in the Live Ball Revived era. The number then nearly tripled to 41 in the home run explosion of the Live Ball Enhanced era.

Chapter 10. The Table Setters

1. The baseball literature has very little on leadoff batters. References were found for 88 leadoff batters, whose career statistics were collected and analyzed as the basis for this section.

2. The six leadoff batters with more than 200 home runs were Bobby Bonds (332), Rickey Henderson (297), Craig Biggio (291), Lou Whitaker (244), Paul Molitor (234), and Brady Anderson (210). Of the 88 leadoff batters in this study, 70 had fewer than 100 home runs. The two leadoff batters with a home run average of more than 3% were Bonds (4.7%) and Anderson (3.2%).

3. Bill James, *The New Bill James Historical Baseball Abstract* (New York: Free Press, 2001), pp. 684 and 685.

4. In his early years, Suzuki was a right fielder, but later on he was moved to center field. I have classified him as a right fielder because he has played in many more games in right field than in center field.

Chapter 11. The Table Clearers

1. Seymour Siwoff et al., *The 1992 Elias Baseball Analyst*, (New York: Fireside, 1992), p. 14.

Chapter 12. Left on Base

1. Seymour Siwoff et al., *The 1991 Elias Baseball Analyst*, (New York: Fireside, 1991), p. 45.

Index

Players are listed individually and are cross-referenced under the team with which they spent the most years. The major league team entries are listed at their current location; earlier locations are given at that entry, and codes indicate at which location the player played longest. The designation "tn" after a page number indicates that the citation is to a table footnote only.

Aaron, Hank: profile, 95, 96; as a black player, 93, 105; home runs compared with other players, 66, 106, 107, 126, 206, 208; Triple Crown, 221; Double Crown, 224; rank in Dead Ball Interval, 94–97; rank among right fielders, 176, 177; rank as a middle-of-the-order batter, 205, 207; dream team, 231

Abreu, Bobby, 176

Advanced weighted measures, 2–4, 7, 9

Albert, Jim, 2, 4, 8, 9, 29

Allen, Dick: profile, 98, 99; as a black player, 105; rank in Dead Ball Interval, 94, 95, 97; rank among first basemen, 162, 163; rank as a middle-of-the-order batter, 205, 207

Alomar, Roberto, 155

Alou, Matty, 185tn

American Association, 25, 33, 34, 36, 37, 235–236n2

Anaheim Angels players. *See* California

Anderson, Brady, 185–187

Anderson, Greg, 120, 121

Andreff, Wladimir, 235n1

Anson, Cap: profile, 32; rank in Era of Constant Change, 28–32; rank among first basemen, 161, 162; rank as a middle-of-the-order batter, 204tn; Double Crown, 224

Aparacio, Luis, 105, 152, 188

Appling, Luke, 149, 153

Arnold, Patrick, 121

Ashburn, Richie, 92, 170, 171, 185, 188

Atlanta (A) / Milwaukee (M) / Boston (B) Braves players. *See* Hank Aaron (M, A), Wally Berger (B), Rico Carty (A, M), Hugh Duffy (B), Rafael Furcal (A), Tommy Holmes (B), Andruw Jones (A), Chipper Jones (A), Javy Lopez (A), Rabbit Maranville (B), Eddie Mathews (M), Tommy McCarthy (B), Billy Nash (B), Lonnie Smith (A), Joe Torre (M, A), Sam Wise (B)

Averill, Earl: profile, 73, 74; rank in Live Ball Era, 63–65; rank among center fielders, 168–170

Bagwell, Jeff, 127, 128, 161, 203

Baines, Harold, 109, 179, 180

Baker, Frank "Home Run": profile, 58; rank in Dead Ball Era, 51, 53, 60; rank as a third baseman, 158, 159; Double Crown, 224tn

Baltimore Orioles players. *See* Brady Anderson, Mark Belanger, Don Buford, Hughie Jennings, Davey Johnson, Joe Kelley, John McGraw, Eddie Murray, Cal Ripken, Jr., Brooks Robinson

Bancroft, Dave, 153, 188

Banks, Ernie: profile, 91, 92; rank in Live Ball Continued Era, 80; rank among shortstops, 149, 151, 152; dream team, 232; mentioned, 104, 163

Baseball Hall of Fame, estimated basis for election of players to: the Era of Constant Change, 35, 36; the Live Ball Interval, 47; the Dead Ball Era, 60; the Live Ball Era, 75, 76; the Live Ball Continued Era, 92; the Dead Ball Interval, 105; the Live Ball Revived Era, 117, 118; shortstops, 152, 153; second basemen, 155, 156; third basemen, 159, 160; first basemen, 163, 164; catchers, 166, 167; center fielders, 169, 171; left fielders, 174; right fielders, 177, 178; summary, 181

Bases per plate appearance (BPPA) explained, 2, 4–6, 9–16, 233, 234

Baylor, Don, 178–180

Beckley, Jake: profile, 45; rank in Live Ball Interval, 40–42; mentioned, 47

Belanger, Mark, 117

Belle, Albert: profile, 133, 134; rank in Live Ball Enhanced Era, 127–129; rank among left fielders, 172, 173; rank as a middle-of-the-order batter, 14, 203, 205, 207; mentioned, 209

Beltran, Carlos, 192

Bench, Johnny: profile, 102, 103; rank in Dead Ball Interval, 97; rank among

catchers, 165–167; Double Crown, 224tn

Bennett, Jay, 2, 4, 8, 9, 29

Berger, Wally, 169tn

Berkman, Lance: profile, 131; rank in Live Ball Enhanced Era, 127–129; rank as a first baseman, 161–163; rank as a middle-of-the-order batter, 207, 208

Berra, Yogi: rank in Live Ball Continued Era, 80; rank among catchers, 165–167

Bescher, Bob, 185tn

Biggio, Craig: profile, 196; rank as a lead-off batter, 185, 187, 189

Bishop, Max, 155, 185

Blue, Lu: profile, 195; rank as a leadoff batter, 185, 187

Boggs, Wade: profile, 190, 191; rank in Live Ball Revived Era, 108; rank among third basemen, 158; rank as a leadoff batter, 15, 184, 185, 187–189, 230; mentioned, 117, 160

Bonds, Barry: profile, 130, 131; home runs, 83, 96, 119, 132, 189, 206, 208; steroids, 120, 121, 123, 125, 126, 128, 132; rank in Live Ball Enhanced Era, 127–129; rank among left fielders, 172, 173; rank as a middle-of-the-order batter, 202–204, 207, 230

Bonds, Bobby: profile, 189; rank as a lead-off batter, 184, 185, 187, 188, 231

Boston Braves players. *See* Atlanta

Boston Red Sox players. *See* Wade Boggs, Jimmy Collins, Joe Cronin, Dom DiMaggio, Bobby Doerr, Nomar Garciaparra, Harry Hooper, Fred Lynn, David Ortiz, Johnny Pesky, Rico Petrocelli, Manny Ramirez, Jerry Remy, Jim Rice, Reggie Smith, Ted Williams, Carl Yastrzemski

Boswell, Tom, 2, 3

Boudreau, Lou: Boudreau Shift, 81; rank in Live Ball Continued Era, 80; rank among shortstops, 149tn, 151, 152

Bresnahan, Roger: profile, 59, 60; rank in Dead Ball Era, 51, 53; rank among catchers, 165tn, 166, 167

Brett, George: profile, 113, 114; rank in Live Ball Revived Era, 107–109, 117; rank among third basemen, 158

Brock, Lou, 105

Brooklyn Dodgers players. *See* Los Angeles

Brouthers, Dan: profile, 23, 30; rank in Era of Constant Change, 28, 29, 31, 31, 35, 229; rank among first basemen, 161, 161, 162; rank as a middle-of-the-order batter, 202–204, 231; Double Crown, 224

Brown, Tom, 24, 185tn

Browning, Pete, 28, 29, 31, 168, 169, 204

Buffalo Bisons players. *See* Dan Brouthers, Hardy Richardson, Jack Rowe, Deacon White

Buford, Don, 185tn

Burkett, Jesse: rank in Live Ball Interval, 39, 40, 47; rank among left fielders, 172–174; rank as a leadoff batter, 184–186, 188, 230

Burns, George Joseph: profile, 195, 196; rank as a leadoff batter, 186, 187

Burns, Thomas "Oyster": profile, 34; rank in Era of Constant Change, 29, 31

Bush, Donnie, 198, 199, 201, 230

Cabrera, Miguel, 158, 159

California / Anaheim / Los Angeles Angels players. *See* Don Baylor, Chili Davis, Brian Downing, David Eckstein, Jim Fregosi, Troy Glaus, Bobby Grich

Camilli, Dolph, 209

Campanella, Roy: profile, 89; compared with other black players, 90, 91, 104; rank in Live Ball Continued Era, 80, 91; rank among catchers, 165–167, 230, 231; dream team, 231

Canseco, Jose: profile, 109, 110; steroids, 110, 137; rank in Live Ball Revived Era, 108, 109; rank as a designated hitter, 178–180

Carew, Rod: profile, 190; rank in Dead Ball Interval, 94, 95, 105; rank as a first baseman, 161, 163; rank as a leadoff batter, 184, 185, 187–189, 230; mentioned, 156

Carey, Max, 76, 170, 171, 188

Carter, Gary, 117, 166

Carty, Rico: profile, 101; rank in Dead Ball Interval, 95, 97, 105

Cash, Norm: profile, 102; rank in Dead Ball Interval, 95, 97

Cedeno, Cesar, 105

Cepeda, Orlando: profile, 101; rank in Dead Ball Interval, 95, 97, 99, 100, 105

Chadwick, Henry, 2

Chance, Frank, 51, 53, 60, 163, 164

Chapman, Ray, 62

Chavez, Eric, 158tn

Chicago Cubs players. *See* Cap Anson, Ernie Banks, Frank Chance, Kiki Cuyler, Bill Dahlen, Abner Dalrymple, Johnny Evers, Augie Galan, George Gore, Stan Hack, Gabby Hartnett, Billy Herman, King Kelly, Bill Nicholson, Fred Pfeffer, Aramis Ramirez, Jimmy Ryan, Ryne Sandberg, Ron Santo, Sammy Sosa, Riggs Stephenson, Joe Tinker, Billy Williams, Ned Williamson, Hack Wilson, Heinie Zimmerman

Chicago White Sox players. *See* Luis Aparacio, Luke Appling, Harold Baines, Eddie Collins, Carlton Fisk, Nellie Fox, Ray Schalk, Frank Thomas

Childs, Cupid: profile, 46, 47; rank in Live Ball Era, 40–42, 47; rank as a middle-of-the-order batter, 204tn

Cincinnati Reds players. *See* Johnny Bench, Bob Bescher, Eric Davis, Adam Dunn, George Foster, Bug Holliday, Dummy Hoy, Ted Kluszewski, Barry

Cincinnati Reds players (*continued*)
Larkin, Ernie Lombardi, Bid McPhee,
Joe Morgan, Heinie Peitz, Tony Perez,
Pokey Reese, John Reilly, Frank Robin-
son, Pete Rose, Edd Roush, Elmer Smith

Clark, Jack: profile, 114; rank in Live Ball
Revived Era, 108, 109; rank among
right fielders, 176, 177

Clark, Will: profile, 111, 112; rank in Live
Ball Revived Era, 108, 109

Clarke, Fred, 50–52, 60

Clemens, Roger: steroids, 122, 123

Clemente, Roberto, 95tn, 105, 178

Clements, Jack, 31, 165, 166

Cleveland Blues player. *See* Fred Dunlap

Cleveland Indians players. *See* Earl Averill,
Albert Belle, Lou Boudreau, Jesse Bur-
kett, Larry Doby, Elmer Flick, Charlie
Hickman, Joe Jackson, Ken Keltner,
Nap Lajoie, Kenny Lofton, Al Rosen,
Joe Sewell, Tris Speaker, Jim Thome,
Andre Thornton, Hal Trosky

Cleveland Spiders players. *See* Cupid
Childs, Ed McKean

Clutch hitting, 215, 216

Cobb, Ty: profile, 49, 52–54; rank in Dead
Ball Era, 50–53, 229, 230; rank among
center fielders, 168–170, 230, 231; rank
as a middle-of-the-order batter, 203,
205, 207, 230; Triple Crown, 221, 222,
225; Double Crown, 224, 225; dream
team, 231; mentioned, 71, 72

Cochrane, Mickey, 64, 76, 164–167,
204tn

Codell, Barry, 3

Collins, Eddie: rank in Dead Ball Era,
50–53, 57; rank among second basemen, 155, 156; rank as a middle-of-the-
order batter, 204

Collins, Jimmy, 60, 160

Colorado Rockies players. *See* Todd Helton, Matt Holliday, Larry Walker

Combs, Earle: profile, 193, 194; rank as a

center fielder, 169–171; rank as a leadoff batter, 185, 187, 188; mentioned, 76

Connor, Roger: profile, 32, 33; rank in Era
of Constant Change, 28–31, 44; rank
among first basemen, 161, 162; rank as a
middle-of-the-order batter, 204tn

Cooper, Walker, 167

Cravath, Gavvy: profile, 55, 56; rank in
Dead Ball Era, 51–53; Triple Crown,
222; Double Crown, 224tn, 225tn

Crawford, Sam: profile, 57, 58; quoted,
49; rank in Dead Ball Era, 51, 53; rank
among right fielders, 176, 177; rank as
a middle-of-the-order batter, 205, 207;
Double Crown, 225tn

Cronin, Joe: profile, 74, 75; rank in Live
Ball Era, 65; rank among shortstops,
149, 151, 152

Cullenbine, Roy, 79tn, 176tn

Cuyler, Kiki, 176tn, 178

Dahlen, Bill, 53

Dalrymple, Abner, 185tn

Daly, Tom, 155tn

Damon, Johnny, 185, 187, 197, 200

Davis, Alvin, 108tn

Davis, Chili, 179, 180

Davis, Eric: profile, 114, 115; rank in Live
Ball Revived Era, 108, 109

Davis, George: rank in Live Ball Interval,
40–42, 47; rank among shortstops, 149,
151, 153

Davis, Spud, 165

Davis, Tommy, 224tn

Dawson, Andre, 117, 118, 178

Delahanty, Ed: profile, 38, 41, 42; rank in
Live Ball Interval, 39–42, 47, 229; rank
among left fielders, 172–174; rank as a
middle-of-the-order batter, 203, 231;
Double Crown, 224tn

Delgado, Carlos, 161tn, 162, 163, 205, 207

Detroit Tigers players. *See* Lu Blue,
Donnie Bush, Norm Cash, Ty Cobb,

Sam Crawford, Roy Cullenbine, Cecil Fielder, Charlie Gehringer, Hank Greenberg, Ned Hanlon, Harry Heilmann, Pinky Higgins, Davey Jones, Al Kaline, George Kell, Ron LeFlore, Marty McIntyre, Germany Schaefer, Mickey Tettleton, Alan Trammell, Bobby Veach, Vic Wertz, Lou Whitaker, Rudy York

Dickey, Bill: profile, 74; rank in the Live Ball Era, 65; rank among catchers, 164–167

DiMaggio, Dom: profile, 194, 195; rank as a leadoff batter, 184, 185, 187; mentioned, 91

DiMaggio, Joe: profile, 81, 82; rank in Live Ball Continued Era, 77–80, 83, 84, 90; rank among center fielders, 168–170; rank as a middle-of-the-order batter, 203, 207, 231; dream team, 232; mentioned, 194, 221

DiMaggio, Vince, 194

Doby, Larry, 92, 167, 170

Doerr, Bobby: profile, 91; rank in Live Ball Continued Era, 80, 90; rank among second basemen, 154–156

Donlin, Mike: profile, 58, 59; rank in Dead Ball Era, 50, 51, 53, 57; rank among center fielders, 169tn, 170

Donovan, Patsy, 42

Double Crown, 223–226; winners, 224

Doubleday, Abner, 23

Downing, Brian, 179

Doyle, Jack, 42

Dream team, 231, 232

Drew, J. D., 176tn

Duffy, Hugh: profile, 43; rank in Live Ball Interval, 39–42; rank among center fielders, 168–171; Triple Crown, 222; Double Crown, 224, 225

Dunlap, Fred, 224, 225

Dunn, Adam, 222

Dykstra, Len, 137, 185

Eckstein, David, 198, 199, 201

Edmonds, Jim, 29, 169, 170

Elliott, Bob, 158, 159

Evers, Johnny, 60, 156, 164

Ewing, Buck: profile, 34, 35; rank in Era of Constant Change, 29, 31, 35; rank among catchers, 165, 166

Fain, Ferris, 161tn, 204tn

Fantasy League Baseball, 233, 234

Farrell, Duke: profile, 47; rank in Live Ball Interval, 42

Fernandez, Tony, 187

Ferrell, Rick, 76, 165tn, 166

Fielder, Cecil, 224tn

Fisk, Carlton, 117, 166, 167

Fletcher, Elbie: profile, 196; rank as a leadoff batter, 185, 187, 230

Flick, Elmer, 57, 60, 178

Florida Marlins players. *See* Miguel Cabrera, Derek Lee, Gary Sheffield

Foster, George, 224tn

Fox, Nellie, 92, 156

Foxx, Jimmie: profile, 68; rank in Live Ball Era, 63–65; rank among first basemen, 161–163; rank as a middle-of-the-order batter, 202–204, 207, 230; Triple Crown, 221, 222, 226; Double Crown, 224tn, 225

Frazee, Harry, 65

Fregosi, Jim, 97

Frisch, Frankie, 76, 155tn, 156

Furcal, Rafael, 149tn

Furillo, Carl, 79tn

Galan, Augie: profile, 189, 190; rank as a leadoff batter, 14, 184, 185, 187

Gamble, Oscar, 179, 180

Garciaparra, Nomar, 127tn, 129, 149–151

Gehrig, Lou: profile, 66, 67; rank in Live Ball Era, 63–65, 68, 71; rank among first basemen, 161–163; rank as a middle-of-the-order batter, 202–204,

Gehrig, Lou (*continued*)
207, 209, 230; Triple Crown, 221, 222,
225; dream team, 231
Gehringer, Charlie, 64tn, 154–156
Gershman, Michael, 61
Giambi, Jason, 127tn, 128
Gibson, Josh, 231
Giles, Brian, 176tn
Glaus, Troy, 137, 158, 159
Gonzalez, Juan, 128, 137
Gordon, Joe, 92, 155, 156
Gordon, Sid: profile, 87; rank in Live Ball
Continued Era, 78–80
Gore, George, 29, 168
Goslin, Goose, 174
Graham, Trevor, 121
Greenberg, Hank: profile, 67, 68; rank in
Live Ball Era, 63–65, 68; rank among
first basemen, 161–163; rank as a
middle-of-the-order batter, 202–204,
207, 209; Double Crown, 224; dream
team, 232; mentioned, 86, 115
Grich, Bobby: profile, 116, 117; rank in
Live Ball Revived Era, 108tn, 109
Griffey, Ken, Jr.: profile, 168; rank in Live
Ball Enhanced Era, 129, 136; rank
among center fielders, 169, 170
Griffin, Mike, 29, 168, 169
Guerrero, Pedro, 108tn
Guerrero, Vladimir: rank in Live Ball En-
hanced Era, 127–129; rank among right
fielders, 175–177; rank as a middle-of-
the-order batter, 203; mentioned, 223
Gwynn, Tony: rank in Live Ball Revived
Era, 108, 114, 117; rank among right
fielders, 176tn, 178; rank as a middle-of-
the-order batter, 204tn

Hack, Stan: profile, 195; rank as a leadoff
batter, 185, 187
Hackenschmidt, George, 70
Hafey, Chick, 76, 172tn
Halberstam, David, 77, 91

Hall, George, 225tn
Hamilton, Billy: profile, 45; rank in Live
Ball Interval, 39, 40, 42, 47; rank among
center fielders, 168, 169, 171; rank as
a leadoff batter, 184–186, 188, 230;
mentioned, 204
Hanlon, Ned, 44, 46
Hartnett, Gabby, 65, 164–167
Hartsel, Topsy: profile, 193; rank among
leadoff batters, 185–187, 230
Hebner, Richie, 97
Heilmann, Harry: profile, 71, 72; rank in
Live Ball Era, 63–65; rank among right
fielders, 175–177; rank as a middle-of-
the-order batter, 204tn, 207
Helton, Todd: profile, 132, 133; rank in
Live Ball Enhanced Era, 127–129; rank
among first basemen, 161–163; rank as
a middle-of-the-order batter, 202, 203,
205, 207; Triple Crown, 222; Double
Crown, 224tn, 226; mentioned, 215
Henderson, Rickey: profile, 194; rank
in Live Ball Revived Era, 117, 118;
rank among leadoff batters, 184, 185,
187–189
Henrich, Tommy: profile, 89, 90; rank in
Live Ball Continued Era, 80
Herman, Babe, 176, 177
Herman, Billy, 76, 155tn, 156
Hernandez, Keith, 98, 108tn
Hickman, Charlie: profile, 59; rank in
Dead Ball Era, 53
Higgins, Pinky, 65
Hines, Paul, 185, 221, 222, 225, 226
Holliday, Bug: profile, 34; rank in Era
of Constant Change, 28, 29, 31; rank
among center fielders, 168, 169tn
Holliday, Matt, 224tn, 226, 227
Holmes, Tommy, 225tn
Hooper, Harry, 60, 178
Hornsby, Rogers: profile, 68, 69; rank
in Live Ball Era, 63–65; rank among
second basemen, 154–156, 230; rank as

a middle-of-the-order batter, 202, 203, 207, 230; Triple Crown, 221, 222, 225, 226, 231; Double Crown, 224, 225; dream team, 231; mentioned, 42, 218

Horowitz, Mikhail, 61

Houston Astros players. *See* Jeff Bagwell, Lance Berkman, Craig Biggio, Cesar Cedeno, Bob Watson, Jimmy Wynn

Howard, Frank: profile, 101, 102; rank in Dead Ball Interval, 95, 96

Howard, Ryan, 224tn, 226, 227

Hoy, Dummy, 40, 185

Hrbek, Kent: profile, 112, 113; rank in Live Ball Revived Era, 108, 109

Huff, Aubrey, 179

Huggins, Miller, 57, 185tn, 188

Jackson, Joe: profile, 54; rank in Dead Ball Era, 50–53; rank among left fielders, 172, 173; rank as a middle-of-the-order batter, 204, 205, 207; dream team, 232

Jackson, Reggie, 108, 117, 126, 178

Jackson, Travis, 153

James, Bill, 2, 3, 144, 184, 202

Jennings, Hughie: profile, 44, 45; rank in Live Ball Interval, 40–42, 46, 47; rank among shortstops, 149–151, 153

Jeter, Derek: profile, 200; rank in Live Ball Enhanced Era, 127tn; rank among shortstops, 149–152; rank among number 2 batters, 15, 197–199, 230

Johnson, Bob, 64tn, 65, 173, 209

Johnson, Cliff, 179, 180

Johnson, Davey, 97

Jones, Andruw, 226

Jones, Chipper: profile, 167; rank in Live Ball Enhanced Era, 127, 129, 160; rank among third basemen, 157–159; rank as a middle-of-the-order batter, 204; dream team, 232

Jones, Davey, 58

Jones, Marion, 121

Joyce, Bill: profile, 42, 43; rank in Live Ball Interval, 39, 40, 42, 47; rank among third basemen, 157, 158; rank as a middle-of-the-order batter, 203

Kaline, Al, 95, 105, 178, 218

Kansas City Athletics players. *See* Oakland

Kansas City Royals players. *See* Carlos Beltran, George Brett, Johnny Damon, Hal McRae, Amos Otis, Mike Sweeney, Danny Tartabull, Willie Wilson

Kaplan, Dave, 228

Keeler, Wee Willie, 51, 57, 60, 178, 204tn

Kell, George, 80, 92, 158tn, 160, 224

Keller, Charlie: profile, 84; rank in Live Ball Continued Era, 78–80; rank among left fielders, 172, 173

Kelley, Joe: profile, 43, 44; rank in Live Ball Interval, 39–42, 47; rank among left fielders, 172–174

Kelly, King, 29, 36, 178, 204tn

Keltner, Ken, 88

Kent, Jeff: profile, 136, 137; rank in Live Ball Enhanced Era, 129; rank among second basemen, 154–156

Killebrew, Harmon: profile, 100; rank in Dead Ball Interval, 94, 95, 97; rank among first basemen, 162, 163; Double Crown, 224tn

Kiner, Ralph: profile, 85, 86; rank in Live Ball Continued Era, 78–80; rank among left fielders, 172, 173; rank as a middle-of-the-order batter, 204tn

Klein, Chuck: rank in Live Ball Era, 64; rank among right fielders, 176, 177, 177; Triple Crown, 221, 222, 225

Kluszewski, Ted, 79tn

Koenig, Mark: profile, 201; rank among number 2 batters, 15, 198, 199, 230

Kruk, John, 108

Lajoie, Nap: profile, 56, 57; rank in Dead Ball Era, 50–53, 57; rank among second

Lajoie, Nap (*continued*)
 basemen, 154–156; rank as a middle-
 of-the-order batter, 204tn, 205, 207;
 Triple Crown, 221, 222, 225; Double
 Crown, 224tn
Landis, Kenesaw Mountain, 54, 62
Lane, Ferdinand, 2, 3
Larkin, Barry, 149, 150, 152
Larkin, Henry: profile, 33, 34; rank in Era
 of Constant Change, 29, 31
Latham, Arlie, 185tn
Lazzeri, Tony, 65, 154–156
Lee, Derek, 225tn
LeFlore, Ron, 185tn
Leonard, Buck, 231
Lindstrom, Freddie, 76, 158tn, 160
Lofton, Kenny, 184–187
Lombardi, Ernie, 165, 167
Lopez, Javy, 165
Loretta, Mark, 198, 199, 230
Los Angeles (LA), Brooklyn (B) Dodgers
 players. *See* Dolph Camilli (B), Roy
 Campanella (B), Tom Daly (B), Tommy
 Davis (LA), Carl Furillo (B), Mike
 Griffin (B), Pedro Guerrero (LA), Babe
 Herman (B), Pee Wee Reese (B), Jackie
 Robinson (B), Eddie Stanky (B), Zack
 Wheat (B)
Luzinski, Greg, 109
Lynn, Fred, 108tn, 109, 113, 225tn
Lyons, Denny: profile, 33; rank in Era
 of Constant Change, 28, 29, 31; rank
 among third basemen, 157, 158; rank as
 a middle-of-the-order batter, 204tn

Madlock, Bill, 108tn, 158tn
Mantle, Mickey: profile, 84, 85; rank in
 Live Ball Continued Era, 78–80, 83,
 84, 86, 87; rank among center fielders,
 168–170; rank as a middle-of-the-order
 batter, 203, 207; Triple Crown, 221,
 222, 225, 226; mentioned, 218
Manush, Heinie, 64tn, 172tn, 174

Maranville, Rabbit, 76, 152
Maris, Roger, 66, 102
Marquard, Rube, 58
Martinez, Edgar, 127, 179, 180, 204tn
Mathews, Eddie: rank in Live Ball Contin-
 ued Era, 79, 80, 88; rank among third
 basemen, 157–159
Mathewson, Christy, 56, 58
Mattingly, Don, 108tn, 194
Mays, Carl, 62
Mays, Willie: profile, 87, 88; rank in Live
 Ball Continued Era, 78–80, 84, 86;
 rank among center fielders, 169, 170;
 mentioned, 104, 126
Mazeroski, Bill, 105, 156
McCarthy, Joe, 82, 92, 182
McCarthy, Tommy, 36, 43, 178, 188
McCovey, Willie: profile, 99, 100; rank
 in Dead Ball Interval, 94, 95, 97, 105;
 rank among first basemen, 163; Double
 Crown, 224tn; mentioned, 126
McGraw, John: profile, 45, 46; rank in
 Live Ball Interval, 39, 40, 42, 44, 47;
 rank among third basemen, 157, 158;
 rank as a leadoff batter, 184–186, 188;
 mentioned, 73
McGwire, Mark: profile, 131, 132;
 steroids, 128, 132; rank in Live Ball
 Enhanced Era, 127–129, 131; rank
 among first basemen, 161–163; rank as
 a middle-of-the-order batter, 203, 205,
 207, 209; mentioned, 70
McIntyre, Marty, 58
McKean, Ed, 149, 151
McNamee, Brian, 122
McPhee, Bid, 24, 31, 35, 155tn, 156
McRae, Hal, 179, 180
Medwick, Joe: rank among left fielders,
 172tn, 173; Triple Crown, 221, 222,
 225, 226
Milwaukee Braves players. *See* Atlanta
Milwaukee Brewers players. *See* Mark
 Loretta, Paul Molitor, Robin Yount

Minnesota Twins players. *See* Rod Carew, Kent Hrbek, Harmon Killebrew, Tony Oliva, Kirby Puckett

Mitchell, Kevin: profile, 110, 111; rank in Live Ball Revived Era, 108, 109

Mitchell Commission, 119, 121–126, 130, 131, 137

Mize, Johnny: profile, 82, 83; rank in Live Ball Continued Era, 78–80; rank among first basemen, 161–163; rank as a middle-of-the-order batter, 203, 207, 209; Double Crown, 224, 225

Molitor, Paul: profile, 191; rank in Live Ball Revived Era, 108, 117; rank among designated hitters, 178–180; rank among leadoff batters, 184, 185, 187–189, 230

Montreal Expos players. *See* Gary Carter, Andre Dawson, Vladimir Guerrero, Tim Raines

Morgan, Joe: profile, 104; rank in Dead Ball Interval, 94, 95, 97, 105; rank among second basemen, 154–156; rank as a number 2 batter, 198, 199, 230

Murray, Eddie, 108tn, 117, 126

Musial, Stan: profile, 83, 84; rank in Live Ball Continued Era, 78–80; rank among left fielders, 171–173; rank as a middle-of-the-order batter, 202, 203, 207; Double Crown, 224tn

Myer, Buddy, 155

Nash, Billy, 31, 158

New York Giants players. *See* San Francisco

New York Mets players. *See* Mike Piazza, Darryl Strawberry

New York Yankees players. *See* Yogi Berra, Earle Combs, Bill Dickey, Joe DiMaggio, Oscar Gamble, Lou Gehrig, Joe Gordon, Tommy Henrich, Derek Jeter, Willie Keeler, Charlie Keller, Mark Koenig, Tony Lazzeri, Mickey Mantle, Roger Maris, Don Mattingly, Jorge Posada, Phil Rizzuto, Alex Rodriguez, Babe Ruth, Alfonso Soriano, Dave Winfield

Nicholson, Bill, 224tn

Oakland (O) / Kansas City (KC) / Philadelphia (P) Athletics players. *See* Frank "Home Run" Baker (P), Max Bishop (P), Jose Canseco (O), Eric Chavez (O), Mickey Cochrane (P), Ferris Fain (P), Jimmie Foxx (P), Jason Giambi (O), Topsy Hartsel (P), Rickey Henderson (O), Reggie Jackson (O), Bob Johnson (P), Henry Larkin (P), Denny Lyons (P), Mark McGwire (O), Tony Phillips (O), Wally Schang (P), Al Simmons (P), Harry Stovey (P), Miguel Tejada (O), Gene Tenace (O)

Oliva, Tony, 95tn, 105

Oliver, Al, 108tn, 224tn

O'Neill, Tip: profile, 33; rank in Era of Constant Change, 28, 29, 31; rank among left fielders, 172, 173; rank as a middle-of-the-order batter, 204tn; Triple Crown, 221–223, 225

O'Rourke, Jim, 29, 35, 174

Ortiz, David: profile, 134, 135; steroids, 123, 128; rank in Live Ball Enhanced Era, 123, 128, 129; rank among designated hitters, 179, 180; rank as a middle-of-the-order batter, 203, 205, 207; Double Crown, 226; strength, skill, and timeliness award, 227

Otis, Amos: profile, 103; rank in Dead Ball Interval, 97

Ott, Mel: profile, 72, 73; rank in Live Ball Era, 63–65; rank among right fielders, 175–177; rank as a middle-of-the-order batter, 203, 207

Palmeiro, Rafael, 125, 126, 137

Palmer, Pete, 2, 3

Peitz, Heinie, 42

Perez, Tony, 105

Pesky, Johnny: profile, 200, 201; rank among shortstops, 149; rank among number 2 batters, 198, 199, 230; mentioned, 91

Petrocelli, Rico: profile, 104; rank in Dead Ball Interval, 97

Pettitte, Andy, 122

Pfeffer, Fred, 31, 35

Philadelphia Athletics players. *See* Oakland

Philadelphia Phillies players. *See* Bobby Abreu, Dick Allen, Richie Ashburn, Dave Bancroft, Jack Clements, Gavvy Cravath, Spud Davis, Ed Delahanty, Len Dykstra, Billy Hamilton, Ryan Howard, Chuck Klein, John Kruk, Greg Luzinski, Scott Rolen, Jimmy Rollins, Juan Samuel, Wally Schang, Mike Schmidt, Roy Thomas, Sam Thompson

Phillips, Tony, 187

Piazza, Mike: profile, 135, 136; rank in Live Ball Enhanced Era, 127, 129; rank among catchers, 164–167; dream team, 232

Pietrusza, David, 61

Pittsburgh Pirates players. *See* Jake Beckley, Max Carey, Fred Clarke, Roberto Clemente, Vince DiMaggio, Patsy Donovan, Bob Elliott, Elbie Fletcher, Richie Hebner, Ralph Kiner, Bill Madlock, Bill Mazeroski, Al Oliver, Manny Sanguillen, Willie Stargell, Pie Traynor, Arky Vaughan, Honus Wagner, Lloyd Waner, Paul Waner, Glenn Wright

Posada, Jorge, 129, 164–167

Potential runs per game (PRG) explained, 16

Providence Grays player. *See* Paul Hines

Puckett, Kirby, 108, 117, 118, 169tn, 170, 171

Pujols, Albert: profile, 128; rank in Live Ball Enhanced Era, 127–129; rank

among first basemen, 161–163; rank as a middle-of-the-order batter, 202–204, 207, 230; Triple Crown, 222, 223; strength, skill, and timeliness award, 227; dream team, 232

Radomski, Kirk, 121, 122

Raines, Tim: profile, 191, 192; rank as a leadoff batter, 184, 185, 187, 188, 230

Ramirez, Aramis, 158, 159

Ramirez, Manny: profile, 128, 129; steroids, 128, 129; rank in Live Ball Enhanced Era, 123, 127–129; rank among left fielders, 172, 173; rank as a middle-of-the-order batter, 203, 205, 207, 208; mentioned, 222

Reese, Pee Wee: profile, 197; rank as a leadoff batter, 187, 188; mentioned, 92, 153

Reese, Pokey, 192

Reilly, John, 222, 225, 226

Remy, Jerry, 198, 199

Rice, Jim: profile, 113; rank in Live Ball Enhanced Era, 108, 109, 117; Double Crown, 224tn, mentioned, 174

Richardson, Hardy, 31, 35

Rickey, Branch, 2, 144

Ripken, Cal, Jr.: profile, 116; rank in Live Ball Revived Era, 109, 117; rank among shortstops, 149tn, 153; mentioned, 66

Ritter, Lawrence S., 198, 236n

Rivers, Mickey, 185tn

Rizzuto, Phil, 92, 152, 188

Roberts, Bip, 185tn

Robinson, Brooks, 105, 107, 160

Robinson, Frank: profile, 96–98; rank in Dead Ball Interval, 94, 95, 97, 105; rank among right fielders, 176, 177; rank as a middle-of-the-order batter, 205, 207; Triple Crown, 221–223, 225; dream team, 232; mentioned, 126

Robinson, Jackie: profile, 90, 91; rank in Live Ball Continued Era, 79, 80; rank among second basemen, 154–156; dream team, 232; mentioned, 104

Rodriguez, Alex: profile, 133; steroids, 128; rank in Live Ball Enhanced Era, 123, 127–129; rank among shortstops, 149–152; rank as a middle-of-the-order batter, 203, 205, 207; Double Crown, 224tn, 225tn, 226; strength, skill, and timeliness award, 227; mentioned, 92, 158, 159, 222

Rodriguez, Ivan, 117, 165

Rolen, Scott, 129, 157–159

Rollins, Jimmy, 149tn

Rose, Pete: profile, 241; rank as a leadoff batter, 185, 187, 189, 230

Rosen, Al: profile, 110, 111; rank in Live Ball Continued Era, 79tn, 80; rank among third basemen, 157–159; Triple Crown, 222; Double Crown, 225tn

Roth, Alan, 2

Roush, Edd, 76, 169tn, 170, 171

Rowe, Jack, 31, 35

Ruth, Babe: profile, 61, 63–66; compared with Ted Williams, 206, 208; rank in Live Ball Era, 63–66, 68, 71, 230; rank among right fielders, 175–177, 230; rank as a middle-of-the-order batter, 14, 202–204, 206, 207, 230; Triple Crown, 220–222; Double Crown, 224, 225; dream team, 231; mentioned, 24, 32, 102, 106, 107, 154, 194

Ryan, Jimmy: rank in Live Ball Interval, 40; rank among leadoff batters, 185, 186, 194, 230; Double Crown, 225

Samuel, Juan: profile, 196; rank as a leadoff batter, 185tn, 187

Sandberg, Ryne, 109, 117, 155tn, 156

San Diego Padres players. *See* Brian Giles, Tony Gwynn, Bip Roberts

San Francisco (SF) / New York Giants (NY) players. *See* Barry Bonds (SF), Bobby Bonds (SF), Roger Bresnahan (NY), George Burns (NY), Orlando Cepeda (SF), Jack Clark (SF), Will Clark (SF), Roger Connor (NY), George Davis (NY), Mike Donlin (NY), Jack Doyle (NY), Buck Ewing (NY), Sid Gordon (NY), Travis Jackson (NY), Jeff Kent (SF, NY), Freddie Lindstrom (NY), Willie Mays (SF, NY), Willie McCovey (SF), Kevin Mitchell (SF, NY), Jim O'Rourke (NY), Mel Ott (NY), Cy Seymour (NY), Bill Terry (NY), Mike Tiernan (NY), George Van Haltren (NY), John Ward (NY), Matt Williams (SF), Ross Youngs (NY)

Sanguillen, Manny, 95tn, 105, 165tn

Santo, Ron: profile, 103; rank in Dead Ball Interval, 97; rank among third basemen, 158, 159

Schaefer, Germany, 197–199, 230

Schalk, Ray, 60, 166

Schang, Wally, 165

Schmidt, Mike: profile, 106–109; rank in Live Ball Revived Era, 107–109, 117, 229, 230; rank among third basemen, 157–159, 230; Double Crown, 224; dream team, 231

Schoendienst, Red, 92, 156, 188

Seattle Mariners players. *See* Alvin Davis, Ken Griffey, Jr., Edgar Martinez, Ichiro Suzuki

Seitzer, Kevin, 108tn

Selig, Bud, 121

Severeid, Hank, 53

Sewell, Joe, 76, 149, 151, 152

Seymour, Cy, 222, 224tn, 225tn

Seymour, Harold, 25tn

Sheffield, Gary, 137, 176, 177

Silverman, Matthew, 61

Simmons, Al: profile, 70, 71; rank in Live Ball Era, 64, 65, 72; rank among left fielders, 172, 173; rank as a middle-of-the-order batter, 203, 207

Simmons, Ted, 109

Sisler, George, 64tn, 204tn, 225tn

Slaughter, Enos, 79tn, 92, 178

Smith, Elmer: profile, 46; rank in Live Ball Interval, 40–42, 47

Smith Lonnie: profile, 192; rank among leadoff batters, 185, 187

Smith, Ozzie, 117, 118, 152

Smith, Reggie, 95

Snider, Duke: profile, 86, 87; rank in Live Ball Continued Era, 79, 80; rank among center fielders, 169, 170

Soriano, Alfonso, 129, 155, 156

Sosa, Sammy, 70, 119, 123, 132, 137

Speaker, Tris: profile, 54, 55; rank in Dead Ball Era, 50–53, 57; rank among center fielders, 168–170; rank as a middle-of-the-order batter, 204; mentioned, 72

Stanky, Eddie, 185

Stargell, Willie: profile, 98; rank in Dead Ball Interval, 95, 97, 105; rank among left fielders, 172tn, 173, 174; rank as a middle-of-the-order batter, 205, 207

Stephens, Vern, 149

Stephenson, Riggs, 64, 172, 204tn

Steroids (performance enhancing drugs). *See* Barry Bonds, Jose Canseco, Roger Clemens, Len Dykstra, Jason Giambi, Troy Glaus, Juan Gonzalez, Mark McGwire, Mitchell Commission, David Ortiz, Rafael Palmeiro, Andy Pettitte, Manny Ramirez, Alex Rodriguez, Gary Sheffield, Sammy Sosa, Miguel Tejada

St. Louis Browns players. *See* Rick Ferrell, Tip O'Neill, Hank Severeid, George Sisler, Vern Stephens, Bobby Wallace, Ken Williams

St. Louis Cardinals players. *See* Lou Brock, Walker Cooper, J. D. Drew, Jim Edmonds, Frankie Frisch, Chick Hafey, Keith Hernandez, Rogers Hornsby, Miller Huggins, Arlie Latham, Joe Medwick, Johnny Mize, Stan Musial,

Albert Pujols, Red Schoendienst, Ted Simmons, Enos Slaughter, Ozzie Smith

Stovey, Harry, 28, 29, 29, 31, 172, 173, 204tn

Strawberry Darryl: profile, 112; rank in Live Ball Revived Era, 107, 108, 109

Streakiness, 216, 217

Strength, skill, and timeliness (SST) award, 227

Suzuki, Ichiro: rank in Live Ball Enhanced Era, 127tn; rank among right fielders, 176; rank as a leadoff batter, 185, 186, 188; mentioned, 204tn, 222

Sweeney, Mike, 179, 180

Szymanski, Stefan, 235n1

Tampa Bay Devil Rays player. *See* Aubrey Huff

Tartabull, Danny: profile, 111; rank in Live Ball Revived Era, 108, 109

Teixeira, Mark, 227

Tejada, Miguel, 137, 149, 151, 152

Tenace, Gene, 95, 97, 165–167

Terry, Bill, 64tn, 204tn

Tettleton, Mickey: profile, 115; rank in Live Ball Revived Era, 109; rank among catchers, 165

Texas Rangers players. *See* Juan Gonzalez, Rafael Palmeiro, Mickey Rivers, Ivan Rodriguez, Mark Teixeira

Thomas, Frank: profile, 131; rank in Live Ball Enhanced Era, 127–129; rank among designated hitters, 179, 180; rank as a middle-of-the-order batter, 203, 205, 207

Thomas, Roy, 185

Thome, Jim: profile, 134; rank in Live Ball Enhanced Era, 127–129; rank among first basemen, 161–163; rank as a middle-of-the-order batter, 203, 205, 207

Thompson, Sam: profile, 30–32; rank in

Era of Constant Change, 28, 29, 31, 35; rank among right fielders, 176, 178; rank as a middle-of-the-order batter, 203; Double Crown, 224tn

Thorn, John, 3

Thornton, Andre, 179, 180

Tiernan, Mike: profile, 44; rank in Live Ball Interval, 39–42, 47; rank among right fielders, 176

Tinker, Joe, 60, 76, 152, 164

Toronto Blue Jays players. *See* Roberto Alomar, Carlos Delgado, Tony Fernandez, Cliff Johnson

Torre, Joe, 165–167, 224tn

Trammell, Alan: profile, 200; rank among number 2 batters, 193, 198, 199, 230

Travis, Cecil, 149tn

Traynor, Pie: profile, 75; rank in Live Ball Era, 65, 75; rank among third basemen, 158–160

Triple Crown, 220–227; winners, 221; near winners, 222

Trosky, Hal, 64tn

Van Haltren, George, 40, 41

Vaughan, Arky: rank in Live Ball Era, 64tn, 65; rank among shortstops, 149–151

Veach, Bobby: profile, 57; rank in Dead Ball Era, 51, 53

Vernon, Mickey, 88

Wagner, Honus: profile, 56; rank in Dead Ball Era, 50–53, 57; rank among shortstops, 149–152, 230; rank as a middle-of-the-order batter, 205, 207; Triple Crown, 222; Double Crown, 224, 225tn, dream team, 231

Walker, Larry: rank in Live Ball Enhanced Era, 127–129; rank among right fielders, 175–177; rank as a middle-of-the-order batter, 203, 207; Double Crown, 223

Wallace, Bobby, 60, 152

Waner, Lloyd: rank among center fielders, 170, 171; rank as a leadoff batter, 185tn, 188; mentioned, 76

Waner, Paul: rank in Live Ball Era, 64tn; rank among right fielders, 176, 178; rank as a middle-of-the-order batter, 204tn; Double Crown, 224tn

Ward, John, 35, 153

Washington Senators players. *See* Duke Farrell, Goose Goslin, Frank Howard, Bill Joyce, Heinie Manush, Buddy Myer, Cecil Travis, Mickey Vernon

Watson, Bob, 95tn

Waxman, Henry, 122

Wertz, Vic, 80, 87, 88

Wheat, Zack, 51, 57, 60, 174

Whitaker, Lou: profile, 193; rank as a leadoff batter, 185, 187, 200, 230

White, Deacon, 158tn, 224tn

Williams, Billy, 95, 105, 174

Williams, Ken, 172

Williams, Matt, 158tn

Williams, Ted: profile, 77, 79–81; compared with Babe Ruth, 206, 208; quoted, 220; rank in Live Ball Continued Era, 78–80, 229, 230; rank among left fielders, 154, 171–173, 230; rank as a middle-of-the-order batter, 202–204, 207, 230; Triple Crown, 221–226, 231; Double Crown, 223–225; dream team, 231; mentioned, 114, 126

Williamson, Ned, 24, 158

Wilson, Hack: profile, 89; rank in Live Ball Era, 64, 65; rank among center fielders, 169, 170; rank as a middle-of-the-order batter, 203, 207

Wilson, Willie, 116

Winfield, Dave, 117, 178

Wise, Sam, 31, 35, 149, 151

Wright, Glenn, 149, 151

Wynn, Jimmy, 97

Yastrzemski, Carl: rank in Dead Ball Interval, 95, 105; Triple Crown, 220–226; mentioned, 174, 218
York, Rudy, 80, 209
Yost, Eddie, 185tn
Young, Cy, 56

Youngs, Ross, 178
Yount, Robin: profile, 115, 116; rank in Live Ball Revived Era, 109, 117; rank among shortstops, 151, 152

Zimmerman, Heinie, 53, 224